Student Solutions Manual
to accompany

For All Practical Purposes
Eighth Edition

Heidi A. Howard
Florida Community College at Jacksonville

W.H. Freeman and Company
New York

© 2010 by W.H. Freeman and Company

ISBN-13: 978-1-4292-2646-2
ISBN-10: 1-4292-2646-3

Printed in the United States of America

First printing

W.H. Freeman and Company
41 Madison Avenue
New York, NY 10010
Houndmills, Basingstoke
RG21 6XS England

www.whfreeman.com

Student Solutions Manual
Table of Contents

PART I: Management Science

PART II: Statistics: The Science of Data

Part III: Voting and Social Choice

PART IV: Fairness and Game Theory

Part V: The Digital Revolution

Part VI: On Size and Growth

Part VII: Your Money and Resources

Chapter 1
Urban Services

Exercise Solutions

1. (a) There are 8 vertices.

   (b) There are 12 edges.

   (c) *A* has valence 3, *B* has valence 2, *C* has valence 3, *D* has valence 2, *E* has valence 4, *F* has valence 4, *G* has valence 3, and *H* has valence 3.

   (d) This is the same as asking if given a vertex, can you reach any other vertex involving at most 2 edges. *A* can reach *B*, *C*, and *F* via one edge. *A* can reach *D*, *E*, *G*, and *H* via two edges. *D* and *F* can similarly reach all other houses in under 20 minutes.

   (e) *B* can reach *A* and *C* in under 10 minutes. *B* can also reach *D* and *F* in under 20 minutes. However, to reach *E*, *G* and *H*, more than 20 minutes is required.

3. (a) This diagram fails to be a graph because a line segment joins a single vertex to itself. The definition being used does not allow this.

   (b) The edge *EC* crosses edges *AD* and *BD* at points which are not vertices; edge *AC* crosses *BD* at a point that is not a vertex.

   (c) This graph has 5 vertices and 6 edges.

5. *E* has valence 0; *A* has valence 1; *H*, *D*, and *G* have valence 2; *B* and *F* have valence 3; *C* has valence 5. *E* is "isolated." *E* might have valence 0 because it is on an island with no road access.

7. (a) *BCGDFB*

   (b) (i) *BD*; *BFD*

      (ii) *CBF*; *CGDF*; *CGDBF*

   (c) *GDBCG*

9. (a) 4 vertices; 4 edges.

   (b) 7 vertices; 6 edges.

   (c) 10 vertices; 14 edges.

11. Remove the edges dotted in the figure below and the remaining graph will be disconnected.

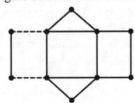

13. (a) 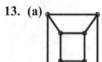 **(b)**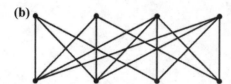

Drawings can vary. Some other graphs could be as follows.

(a) 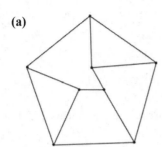 **(b)**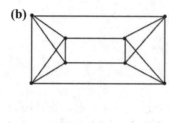

(c) Yes. The sum of the valences of a graph with 8 vertices each of valence 2 is 16. Thus, all such graphs have 8 edges.

15. (a)

(b)

17. (a)  **(b)**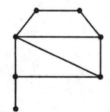

19. The supervisor is not satisfied because all of the edges are not traveled upon by the postal worker. The worker is unhappy because the end of the worker's route wasn't the same point as where the worker began. The original job description is unrealistic because there is no Euler circuit in the graph.

21. There is such an efficient route. The appropriate graph model has an additional edge joining the same pair of vertices for each of the edges shown in the graph of Exercise 19. Since this graph is connected and even-valent, it has an Euler circuit, any one of which will provide a route for the snowplow. Routes without 180-degree turns are better choices.

23.

25. (a) The largest number of such paths is 3. One set of such paths is *AHF*, *ABEF*, and *AGCF*.

(b) This task is simplified by noticing there are many symmetries in this graph. You may notice that starting with *A*, there are only three directions to go. Any number of paths greater than 3 would involve repeating an edge starting from *A*.

(c) In a communication system such a graph offers redundant ways to get messages between pairs of points even when the failure of some of the communication links (edges removed) occurs.

27. Do not choose edge 2, but edges 1 or 10 could be chosen.

29.

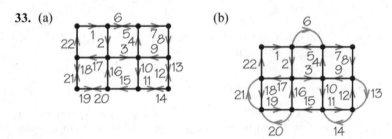

31. Two edges need to be dropped to produce a graph with an Euler circuit. Persons who parked along these stretches of sidewalk without putting coins in the meters would not need to fear that they would get tickets.

33. (a)   **(b)**

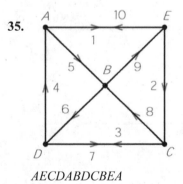

No. Five is the minimum number of edges that must be reused. Fewer than 5 reused edges cannot be achieved.

35.

AECDABDCBEA

37. (a)

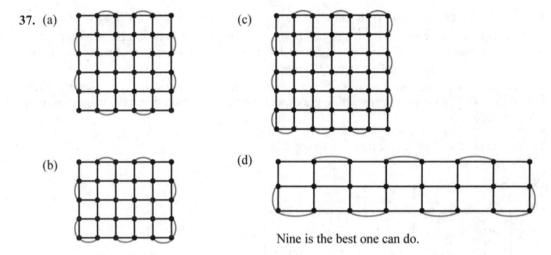

(b)

(d)

Nine is the best one can do.

39. (a) There are four 3-valent vertices. By properly removing two edges adjacent to these four vertices, (edge between left two 3-valent vertices and edge between right two 3-valent vertices) one can make the graph even-valent.

(b) Yes, because the resulting graph is connected and even-valent.

(c) It is possible to remove two edges and have the resulting graph be even-valent.

(d) No, because the resulting graph is not connected, even though it is even-valent.

41. There are many different circuits which will involve three reuses of edges. These are the edges which join up the six 3-valent vertices in pairs.

43. The following graph satisfies the condition. You would eulerize the graph by duplicating each edge exactly once. The two end vertices are odd-valent.

45. There are many circuits that achieve a length of 44,000 feet. The number of edges reused is eight because a shorter length tour can be found by repeating more shorter edges than fewer longer edges.

47. Both graphs (b) and (c) have Euler circuits. The valences of all of the vertices in (a) are odd, which makes it impossible to have an Euler circuit there.

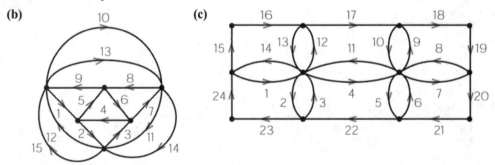

49. If the graph G is connected, the newly constructed graph will be even-valent and, thus, will have an Euler circuit. If G is not connected, the new graph will not have an Euler circuit because it, too, will not be connected.

51. (a)

(b) The best eulerization for the four-circle, four-ray case adds two edges.

(c) Hint: Consider the cases where r is even and odd separately.

53. A graph with six vertices where each vertex is joined to every other vertex will have valence 5 for each vertex.

55. When you attach a new edge to an existing graph, it gets attached at two ends. At each of its ends, it makes the valence of the existing vertex go up by one. Thus the increase in the sum of the valences is two. Therefore, if the graph had an even sum of the valences before, it still does, and if its valence sum was odd before, it still is.

57.

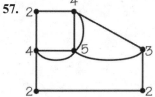

The graph is connected.

59. In chemistry when we say, for example, that hydrogen has valence 1, we mean that it forms one chemical bond with other elements. This usage has similarities with the graph theory concept of valence.

61. A tour that begins and ends at vertex A and which respects the traffic directions would be: *ABDEFBEBFEDBACDCA*. The cutting machine has to make "sharp turns" at some intersections.

Chapter 2
Business Efficiency

Exercise Solutions

1. **(a)** $X_5X_6X_1X_3X_4X_2X_5$

   (b) $X_5X_4X_3X_2X_1X_6X_7X_8X_9X_{10}X_{11}X_{12}X_5$

   (c) $X_5X_4X_3X_1X_2X_7X_6X_9X_8X_5$

3. Yes, for all three graphs.

5. **(a)** A Hamiltonian circuit will remain for (b), but there will be no Hamiltonian circuit for (a), (c).
   (b) Removing an edge between two vertices in a communications network indicates that it would no longer be possible to send messages between these two sites.

7. Other Hamiltonian circuits include *ABIGDCEFHA* and *ABDCEFGIHA*.

9. **(a)** a. Add edge *AB*.

      b. Add edge X_1X_3.

   (b)

   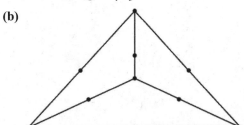

   (c) Add edges X_2X_8, X_8X_6, X_6X_4, and X_4X_2.

11. The graph below has no Hamiltonian circuit and every vertex of the graph has valence 3.

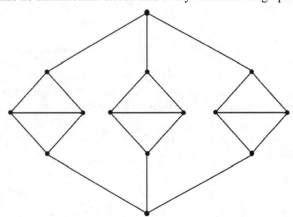

13. (a) Any Hamiltonian circuit would have to use both edges at the vertices X_5, X_4, and X_2. This would cause a problem in the way a Hamiltonian circuit could visit vertex X_1. Thus, no Hamiltonian circuit exists.

   (b) If there were a Hamiltonian circuit, it would have to use the edges X_4 and X_5 and X_6 and X_7. This would make it impossible for the Hamiltonian circuit to visit X_8 and X_9. Thus, no Hamiltonian circuit exists.

15. (a) Yes there is a Hamilton circuit.

   (b) No Hamilton circuit.

   (c) No Hamilton circuit.

17. (a) There is a Hamiltonian path from X_3 to X_4.

   (b) No. There is a Hamiltonian path from X_1 to X_8 in graph (b).

   (c) Here are two examples. A worker who inspects sewers may start at one garage at the start of the work day but may have to report to a different garage for the afternoon shift. A school bus may start at a bus garage and then pick up students to take them to school, where the bus sits until the end of the school day.

19. (a) Hamiltonian circuit, yes. One example is: $X_1X_3X_7X_5X_6X_8X_2X_4X_1$; Euler circuit, yes.

   (b) Hamiltonian circuit, yes. Euler circuit, yes.

   (c) Hamiltonian circuit, yes. Euler circuit, no.

   (d) Hamiltonian circuit, no. Euler circuit, yes. One example is as follows.
$$U_1U_2U_5U_6U_{16}U_{15}U_{11}U_4U_5U_{12}U_{11}U_{10}U_{14}U_{13}U_7U_3U_8U_{10}U_9U_3U_1.$$

21. (a) Hamiltonian circuit, yes; Euler circuit, no.

   (b) Hamiltonian circuit, yes; Euler circuit, no.

   (c) Hamiltonian circuit, yes; Euler circuit, no.

   (d) Hamiltonian circuit, no; Euler circuit, no.

23. (a)

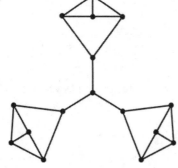

(b)

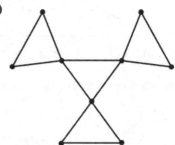

   (c) A graph has an Eulerian path if for two different vertices u and v of the graph there is a path from u to v that uses each edge of the graph once and only once.

25. (a) $7 \times 6 \times 5 \times 4 \times 3 = 2520$

   (b) $7 \times 7 \times 7 \times 7 \times 7 = 16,807$

   (c) $7^5 - 7 = 16,800$

27. (a) $9 \cdot 8 \cdot 7 \cdot 6 \cdot 5 = 15,120$

   (b) $(26)(26)(26) = 17,576$

29. The new system is an improvement since it codes 676 locations compared with 504 for the old system. This is 172 more locations.

31. (a) $(26)(26)(26)(10)(10)(10) - (26)(26)(26) = 26^3 (10^3 - 1) = 17,558,424$

    (b) Answers will vary.

33. With no other restrictions, $10^7 = 10,000,000$. With no other restrictions, $9 \times 10^2 = 900$.

35. These graphs have 6, 10, and 15 edges, respectively. The n-vertex complete graph has $\dfrac{n(n-1)}{2}$ edges. The number of TSP tours is 3, 12, and 60, respectively.

37. (a)

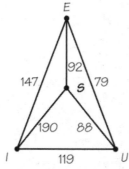

    (b) (1) *UISEU;* mileage $= 119 + 190 + 92 + 79 = 480$
         (2) *USIEU;* mileage $= 88 + 190 + 147 + 79 = 504$
         (3) *UIESU;* mileage $= 119 + 147 + 92 + 88 = 446$

    (c) *UIESU* (Tour 3)

    (d) No.

    (e) Starting from *U*, one gets *UESIU* Tour 1. From *S* one gets *SUEIS* Tour 2; from *E* one gets *EUSIE* Tour 2; and from *I* one gets *IUESI* Tour 1.

    (f) *EUSIE* Tour 2. No.

39. *FMCRF* gets her home in 40 minutes.

41. *MACBM* takes 344 minutes to traverse.

43. A traveling salesman problem.

45. A sewer drain inspection route at corners involves finding a Hamiltonian circuit, and there is such a circuit. If the drains are along the blocks, a route in this case involves solving a Chinese postman problem. Since there are 18 odd-valent vertices, an optimal route would require at least 9 reuses of edges. There are many such routes that achieve 9 reuses.

47. The complete graph shown has a different nearest-neighbor tour that starts at A (*AEDBCA*), a sorted-edges tour (*AEDCBA*), and a cheaper tour (*ADBECA*).

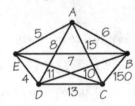

49. The optimal tour is the same but its cost is now $4200 + 10(50) = 4700$.

51. (a) a.   Not a tree because there is a circuit. Also, the wiggled edges do not include all vertices of the graph.
   b.   The circuit does not include all the vertices of the graph.
   (b) a.   The tree does not include all vertices of the graph.
   b.   Not a circuit.
   (c) a.   Not a tree.
   b.   Not a circuit.
   (d) a.   Not a tree.
   b.   Not a circuit.

53. (a) 1, 2, 3, 4, 5, 8; cost is 23
   (b) 1, 1, 1, 2, 2, 3, 3, 4, 5, 6, 6; cost is 34
   (c) 1, 1, 1, 2, 2, 2, 2, 2, 3, 3, 3, 3, 4, 4, 4, 5, 5, 6, 7; cost is 60
   (d) 1, 2, 2, 3, 3, 3, 4, 5, 5, 5, 6, 6; cost is 45

55. The spanning tree will have 27 vertices. H also will have 27 vertices. The exact number of edges in H cannot be determined, but H has at least 26 edges.

57. Yes.

59. Yes. Change all the weights to negative numbers and apply Kruskal's algorithm. The resulting tree works, and the maximum cost is the negative of the answer you get. If the numbers on the edges represent subsidies for using the edges, one might be interested in finding a maximum-cost spanning tree.

61. A negative weight on an edge is conceivable, perhaps a subsidization payment. Kruskal's algorithm would still apply.

63. Three different trees with the same cost are shown:

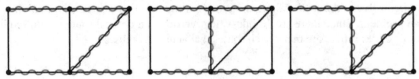

65. (a) True
   (b) False (unless all the edges of the graph have the same weight)
   (c) True
   (d) False
   (e) False

67. (a) Answers will vary for each edge, but the reason it is possible to find such trees is that each edge is an edge of some circuit.

(b) The number of edges in every spanning tree is five, one less than the number of vertices in the graph.

(c) Every spanning tree must include the edge joining vertices C and D, since this edge does not belong to any (simple) circuit in the graph.

69.

	A	B	C	D
A	0	16	13	5
B	16	0	19	11
C	13	19	0	8
D	5	11	8	0

71. (a) The earliest completion time is 22 since the longest path, the unique critical path $T_3T_2T_5$, has length 22.

(b) The earliest completion time is 30 since the longest path, the unique critical path $T_3T_5T_7$, has length 30.

73. The only tasks which if shortened will reduce the earliest completion time are those on the critical path, so in this case, these are the tasks T_1, T_5, and T_7. If T_5 is shortened to 7, then the longest path will have length 28, and this becomes the earliest completion time. The tasks on this critical path are T_1, T_4, and T_7.

75. Different contractors will have different times and order-requirement digraphs. However, in any sensible order requirement digraph, the laying of the foundation will come before the erection of the side walls and the roof. The fastest time for completing all the tasks will be the length of the longest path in the order-requirement digraph.

77. One example is given below.

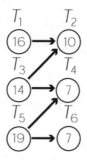

Chapter 3
Planning and Scheduling

Exercise Solutions

1. **(a)** Scheduling final examinations, the times for courses being offered, cleaning of classrooms, etc.

    (b) Scheduling nurses, doctors, operating rooms, medical imaging, etc.

    (c) Scheduling of tracks for departing trains and for arriving trains; schedules for people to man the information desks; schedules for people to sell tickets; etc.

    (d) Scheduling ground and squad car patrols around the clock, officers on surveillance assignments, general work involving nonemergency interaction with the public; arrangements must be made regarding who is on call for emergency duty.

    (e) Scheduling store personnel, inventory work, shelving new books, and restoring books to proper order on the shelves.

    (f) Scheduling of hours the cafe will be open; scheduling of maintenance of the machines used; scheduling employees to operate the cafe; etc.

    (g) Scheduling maintenance and repair of hoses and equipment, personnel to respond to alarms.

    (h) Scheduling of times to produce shows to be broadcast at a later time; scheduling of people with expertise to use the cameras; scheduling of janitorial staff; etc.

3. Some tasks include making sure you have enough folding chairs for people to sit on, ordering the food that will be served, cooking the food to be served, having some takeout food delivered, cleaning prior to the arrival of your guests, etc. The processors used include people, and perhaps the stove, oven, and microwave. Many of the tasks can be going on at the same time.

5. **(a)** Processor 1: T_1, T_2, T_3, T_5, T_7.
    Processor 2: Idle 0 to 2, T_4, T_6, idle 4 to 5.

    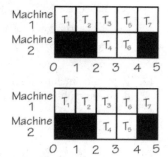

    (b) Processor 1: T_1, T_2, T_3, T_6, T_7.
    Processor 2: Idle 0 to 2, T_4, T_5, idle 4 to 5.

    (c) Yes.
    (d) No.
    (e) T_3 and T_5.

7. **(a)** i. Processor 1: T_1 from 0 to 13, T_3 from 13 to 25, T_6 from 25 to 45.

Processor 2: T_2 from 0 to 18, T_4 from 18 to 27, T_5 from 27 to 35, idle from 35 to 45.

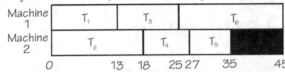

ii. Processor 1: T_1 from 0 to 13, T_3 from 13 to 25, T_4 from 25 to 34, T_5 from 34 to 42.

Processor 2: T_2 from 0 to 18, T_6 from 18 to 38, idle from 38 to 42.

(b) The schedule produced in (ii) is optimal, because the sum of the task times is 80 and no set of tasks can be arranged that will feasibly sum to 40 on each processor.

(c) The critical path is T_2, T_6, and it has length 38. No schedule can be completed by time 38 on two processors because the sum of the task times divided by 2 is 40.

9. **(a)** Yes. Suppose the original order requirement digraph has three tasks each taking 6 time units and no edges. Drawing an edge between two of these tasks now lengthens the critical path to 12.

(b) No. All the pre-existing paths have the same length and some new paths might be created. Thus, the length of the longest path cannot go down.

11. **(a)** Processor 1: T_1, T_6, idle 15 to 21, T_7, idle 27 to 31.

Processor 2: T_2, T_5, T_8.

Processor 3: T_3, T_4, idle from 13 to 31.

(b) Processor 1: T_1, T_6, idle 15 to 21, T_7, idle 27 to 31.

Processor 2: T_3, T_4, idle from 13 to 21, T_8.

Processor 3: T_2, T_5, idle from 21 to 31.

(c) Processor 1: T_4, idle 10 to 11, T_6, idle 18 to 21, T_8.

Processor 2: T_2, T_5, T_7, idle 27 to 31.

Processor 3: T_1, T_3, idle 11 to 31.

13. Examples include inserting identical mirror systems on different models of cars and vaccinating different children against polio.

15. T_1, T_2, T_3, T_4, T_8, T_9, T_{10}, T_{11}, T_5, T_6, T_7, T_{12}.

17. **(a)** No. Consider the tasks that begin after the stretch where all machines are idle. Pick one of these tasks T and say machine 1 was the machine that it was given to. This task was ready for machine 1 just prior to when it began T because no task was just being completed on any other machine at this time because they were all idle. Thus, T should have begun earlier on machine 1.

(b) This schedule cannot arise using the list-processing algorithm, because T_2 should have been scheduled at time 0.

(c) Use the digraph with no edges and the list: T_2, T_1, T_3, T_4, T_5.

19. (a) T_1, T_2, T_3, and T_6 are ready at time 0.

(b) No tasks require that T_1 and T_6 be done before these other tasks can begin.

(c) The critical path consists only of T_6 and has length 20.

(d) Processor 1: T_1, T_6: Processor 2: T_2, T_4, idle from 18 to 30: Processor 3: T_3, T_5, idle from 12 to 30.

(e) No.

(f) Processor 1: T_6, idle from 20 to 22: Processor 2: T_3, T_5, T_1: Processor 3: T_2, T_4, idle from 18 to 22.

(g) Yes.

(h) Another list leading to the same optimal schedule is T_6, T_3, T_2, T_4, T_5, T_1.

21. (a) $5! = 120$

(b) No. Whatever list is used, T_1 must be assigned to the first machine at time 0 because it is the only task ready at time 0.

(c) No. First, while Processor 1 works on T_1. Processor 2 must be idle. Second, the task times are integers with sum 31. If there are two processors, one of the processors must have idle time since when 2 divides 31, there is a remainder of 1.

(d) No.

23. Using the order-requirement digraph shown and any list with one or more processors yields the same schedule:

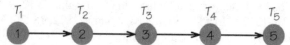

25. (a) One reasonable possibility is (time in min):

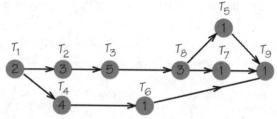

The earliest completion time is 15.

(b) The decreasing-time list is T_3, T_4, T_2, T_8, T_1, T_5, T_6, T_7, T_9. The schedule is Processor 1: T_1, T_4, T_6, idle 7 to 10, T_8, T_5, T_9; Processor 2: idle 0 to 2, T_2, T_3, idle 10 to 13, T_7, idle 14 to 15.

27. No. At time 11, T_4 should been assigned to Machine 1 because Machine 1 was free at this time and T_4 was ready.

29. (a) Task times: $T_1 = 3$, $T_2 = 3$, $T_3 = 2$, $T_4 = 3$, $T_5 = 3$, $T_6 = 4$, $T_7 = 5$, $T_8 = 3$, $T_9 = 2$, $T_{10} = 1$, $T_{11} = 1$, and $T_{12} = 3$. This schedule would be produced from the list: T_1, T_3, T_2, T_5, T_4, T_6, T_7, T_8, T_{11}, T_{12}, T_9, T_{10}.

(b) Task times: $T_1 = 3$, $T_2 = 3$, $T_3 = 3$, $T_4 = 2$, $T_5 = 2$, $T_6 = 4$, $T_7 = 3$, $T_8 = 5$, $T_9 = 8$, $T_{10} = 4$, $T_{11} = 7$, $T_{12} = 9$, and $T_{13} = 3$. This schedule would be produced from the list: T_1, T_5, T_7, T_4, T_3, T_6, T_{11}, T_8, T_{12}, T_9, T_2, T_{10}, T_{13}.

31. (a) (i) Processor 1: T_1, T_3, T_5, T_7, idle from 16 to 20; Processor 2: T_2, T_4, T_6, T_8.

(ii) Processor 1: T_8, T_5, T_4, T_1; Processor 2: T_7, T_6, T_3, T_2.

(b) The schedule in (ii) is optimal.

33. Such criteria include decreasing length of the times of the tasks, order of size of financial gains when each task is finished, and increasing length of the times of the tasks.

35. (a) Machine 1: T_1, T_6, T_{10}, idle from 8 to 9; Machine 2: T_3, T_4, T_{11}, T_{12}; Machine 3: T_2, T_7, idle from 8 to 9; Machine 4: T_5, T_8, T_9, idle from 8 to 9.

(b) Machine 1: T_1, T_8, T_{10}; Machine 2: T_5, T_6, T_2, T_{13}; Machine 3: T_7, T_{12}; Machine 4: T_4, T_{11}, idle from 9 to 12; Machine 5: T_3, T_9, idle from 11 to 12.

The schedule in part (a) had to have idle time because the total task time was 33, and 33 is not exactly divisible by 4. The schedule in part (b) had to have idle time because the total task time was 56, and 56 is not exactly divisible by 5.

37. (a) List for (i) yields (with items coded by task time): Machine 1: 12, 9, 15, idle from 36 to 50; Machine 2: 7, 10, 13, 20. List for (ii) yields: Machine 1: 12, 13, 20; Machine 2: 7, 9, 15, 10, idle from 41 to 45. List for (iii) yields: Machine 1: 20, 12, 9, idle from 41 to 45; Machine 2: 15, 13, 10, 7.

(b) These schedules complete earlier than those where precedence constraints hold. An optimal schedule is possible, however: Machine 1: 20, 10, 13; Machine 2: 12, 15, 9, 7. The associated list is: T_6, T_1, T_5, T_7, T_2, T_4, T_3.

(c) The critical path list is T_6, T_5, T_4, T_1, T_7, T_2, T_3 using the first processor. It finishes at time 41 and is idle until 45, when processor 2 finishes.

39. (a) With this list the order of the tasks on the processors is: Machine 1: 19, 20, 1, 2, 3, 5, 11, 17, 18, 17, 16 (completion time 129); Machine 2: 20, 1, 2, 3, 5, 11, 17, 18, 2, 16, 2, idle 116 to 129. The completion time is 129.

(b) The list is: 20, 20, 19, 19, 18, 18, 17, 17, 17, 16, 16, 11, 11, 5, 5, 3, 3, 2, 2, 2, 1, 1. On two processors we get the schedule: Machine 1: 20, 19, 18, 17, 17, 11, 11, 3, 3, 2, 1, 1 (completion time 123); Machine 2: 20, 19, 18, 17, 16, 16, 5, 5, 2, 2, 2, idle 122 to 123. The completion time is 123 on both machines.

(c) The answer in (b), decreasing-time-list algorithm, is one optimal schedule

41. (a) The tasks are scheduled on the machines as follows: Processor 1: 12, 13, 45, 34, 63, 43, 16, idle 226 to 298; Processor 2: 23, 24, 23, 53, 25, 74, 76; Processor 3: 32, 23, 14, 21, 18, 47, 23, 43, 16, idle 237 to 298.

   (b) The tasks are scheduled on the machines as follows: Processor 1: 12, 24, 14, 34, 25, 23, 16, 16, 76; Processor 2: 23, 23, 21, 63, 43, idle 173 to 240; Processor 3: 32, 23, 53, 74, idle 182 to 240; Processor 4: 13, 45, 18, 47, 43, idle 166 to 240.

   (c) The decreasing-time list is 76, 74, 63, 53, 47, 45, 43, 43, 34, 32, 25, 24, 23, 23, 23, 23, 21, 18, 16, 16, 14, 13, 12. The tasks are scheduled on three machines as follows: Processor 1: 76, 45, 43, 24, 23, 18, 16, 13; Processor 2: 74, 47, 34, 32, 23, 21, 14, 12, idle 257 to 258; Processor 3: 63, 53, 43, 25, 23, 23, 16, idle 246 to 248. The tasks are scheduled on four machines as follows: Processor 1: 76, 43, 24, 23, 16, idle 182 to 194; Processor 2: 74, 43, 25, 23, 16, 13; Processor 3: 63, 45, 32, 23, 18, 12, idle 193 to 194; Processor 4: 53, 47, 34, 23, 21, 14, idle 192 to 194.

   (d) The new decreasing time list is 84, 82, 71, 61, 55, 45, 43, 43, 34, 32, 25, 24, 23, 23, 23, 23, 21, 18, 16, 16, 14, 13, 12. The tasks are scheduled as follows: Processor 1: 84, 45, 43, 25, 23, 23, 16, 12; Processor 2: 82, 55, 34, 32, 23, 18, 14, 13; Processor 3: 71, 61, 43, 24, 23, 21, 16, idle 259 to 271.

43. Examples include jobs in a videotape copying shop, data entry tasks in a computer system, and scheduling nonemergency operations in an operating room. These situations may have tasks with different priorities, but there is no physical reason for the tasks not to be independent, as would be the case with putting on a roof before a house had walls erected.

45. Each task heads a path of length equal to the time to do that task.

47. The times to photocopy the manuscripts, in decreasing order, are 120, 96, 96, 88, 80, 76, 64, 64, 60, 60, 56, 48, 40, 32. Packing these in bins of size 120 yields Bin 1: 120; Bin 2: 96; Bin 3: 96; Bin 4: 88, 32; Bin 5: 80, 40; Bin 6: 76; Bin 7: 64, 56; Bin 8: 64, 48; Bin 9: 60, 60. Nine photocopy machines are needed to finish within 2 minutes using FFD. The number of bins would not change, but the placement of the items in the bins would differ for worst-fit decreasing.

49. (a) Using the next-fit algorithm, the bins are filled as follows: Bin 1: 12, 15; Bin 2: 16, 12; Bin 3: 9, 11, 15; Bin 4: 17, 12; Bin 5: 14, 17; Bin 6: 18; Bin 7: 19; Bin 8: 21; Bin 9: 31; Bin 10: 7, 21; Bin 11: 9, 23; Bin 12: 24; Bin 13: 15, 16; Bin 14: 12, 9, 8; Bin 15: 27; Bin 16: 22; Bin 17: 18.

   (b) The decreasing list is 31, 27, 24, 23, 22, 21, 21, 19, 18, 18, 17, 17, 16, 16, 15, 15, 15, 14, 12, 12, 12, 12, 11, 9, 9, 9, 8, 7. The next-fit decreasing schedule is Bin 1: 31; Bin 2: 27; Bin 3: 24; Bin 4: 23; Bin 5: 22; Bin 6: 21; Bin 7: 21; Bin 8: 19; Bin 9: 18, 18; Bin 10: 17, 17; Bin 11: 16, 16; Bin 12: 15, 15; Bin 13: 15, 14; Bin 14: 12, 12, 12; Bin 15: 12, 11, 9; Bin 16: 9, 9, 8, 7.

   (c) The worst-fit schedule using the original list is Bin 1: 12, 15, 9; Bin 2: 16, 12; Bin 3: 11, 15; Bin 4: 17, 12; Bin 5: 14, 17; Bin 6: 18, 7; Bin 7: 19, 9; Bin 8: 21, 15; Bin 9: 31; Bin 10: 21, 9; Bin 11: 23, 8; Bin 12: 24; Bin 13: 16, 12; Bin 14: 27; Bin 15: 22; Bin 16: 18.

   (d) The worst-fit decreasing schedule would be Bin 1: 31; Bin 2: 27, 9; Bin 3: 24, 12; Bin 4: 23, 12; Bin 5: 22, 14; Bin 6: 21, 15; Bin 7: 21, 15; Bin 8: 19, 17; Bin 9: 18, 18; Bin 10: 17, 16; Bin 11: 16, 15; Bin 12: 12, 12, 11; Bin 13: 9, 9, 8, 7.

51. The bins have a capacity of 120. (First fit): Bin 1: 63, 32, 11; Bin 2: 19, 24, 64; Bin 3: 87, 27; Bin 4: 36, 42; Bin 5: 63. This schedule would take five station breaks; however, the total time for the breaks is under 8 minutes. The decreasing list is 87, 64, 63, 63, 42, 36, 32, 27, 24, 19, 11. (First-fit decreasing): Bin 1: 87, 32; Bin 2: 64, 42, 11; Bin 3: 63, 36, 19; Bin 4: 63, 27, 24. This solution uses only four station breaks.

53. (a) There are theoretical results that show that best fit "usually" performs better than worst fit.

(b) Try 8, 7, 5, 3, 3, 2 in bins of capacity 14.

55. The total performance time exceeds what will fit on four disks. Using first-fit decreasing, one can fit the music on five disks.

57. For problems with few weights to be packed, small integer weights, and a small integer as bin capacity, this method can work well. However, when these special conditions are not met, it is very time-consuming to carry out this method. For example, imagine trying to use this method for 2000 random real numbers of the form $.xyz$, where x, y, and z are decimal digits and the bin capacity is 1.

59. It makes sense to leave bins open as more items arrive to be packed if the cost of having many bins open at once is reasonable and there is room to have many partially-filled bins open without incurring great inconvenience or cost. One such example might be a company that has room for many identical trucks to park as they are loaded with goods to be delivered. There may be complex cost trade-offs between sending off fewer trucks because we wait to pack as much into each truck as possible and sending out more partially-filled trucks.

61. (a) The schedule with four secretaries is as follows: Processor 1: 25, 36, 15, 15, 19, 15, 27; Processor 2: 18, 32, 18, 31, 30, 18; Processor 3: 13, 30, 17, 12, 18, 16, 16, 16, 14; Processor 4: 19, 12, 25, 26, 18, 12, 24, 9. Completion time is 152 minutes.

The schedule with five secretaries is as follows: Processor 1: 25, 25, 31, 12, 16, 14; Processor 2: 18, 12, 17, 12, 15, 30, 9; Processor 3: 13, 32, 26, 16, 15, 18; Processor 4: 19, 36, 18, 19, 24; Processor 5: 30, 18, 15, 18, 16, 27. Completion time is 124 minutes.

(b) The decreasing time list is 36, 32, 31, 30, 30, 27, 26, 25, 25, 24, 19, 19, 18, 18, 18, 18, 18, 17, 16, 16, 16, 15, 15, 15, 14, 13, 12, 12, 12, 9.

The schedule using this list on four processors would be Processor 1: 36, 25, 19, 18, 17, 16, 13, 9; Processor 2: 32, 26, 25, 18, 16, 15, 12; Processor 3: 31, 27, 24, 18, 16, 15, 12, 12; Processor 4: 30, 30, 19, 18, 18, 15, 14. Completion time is 155 minutes.

The schedule using this list on five processors would be Processor 1: 36, 24, 18, 16, 14, 12; Processor 2: 32, 25, 18, 18, 15, 12; Processor 3: 31, 25, 19, 18, 15, 9; Processor 4: 30, 27, 18, 17, 15, 13; Processor 5: 30, 26, 19, 16, 16, 12. Completion time is 120 minutes.

(c) The five-processor decreasing-time schedule is optimal (time 120), but the four-processor decreasing-time schedule is not. One can see this, since when the task of length 17 scheduled on processor 1 and the task of length 18 on processor 3 are interchanged, the completion time is reduced to 154 from 155 for the four-processor, decreasing-time schedule.

(d) As a bin-packing problem, each bin will have a capacity of 60. Using the decreasing list, we obtain the following packings: (First-fit decreasing): Bin 1: 36, 24; Bin 2: 32, 27; Bin 3: 31, 26; Bin 4: 30, 30; Bin 5: 25, 25, 9; Bin 6: 19, 19, 18; Bin 7: 18, 18, 18; Bin 8: 18, 17, 16; Bin 9: 16, 16, 15, 13; Bin 10: 15, 15, 14, 12; Bin 11: 12, 12.

(e) NFD uses 13 bins. Bin 1: 36; Bin 2: 32; Bin 3: 31; Bin 4: 30, 30; Bin 5: 27, 26; Bin 6: 25, 25; Bin 7: 24, 19; Bin 8: 19, 18, 18; Bin 9: 18, 18, 18; Bin 10: 17, 16, 16; Bin 11: 16, 15, 15; Bin 12: 15, 14, 13, 12; Bin 13: 12, 12, 9. WFD uses 11 bins. Bin 1: 36, 24; Bin 2: 32, 26; Bin 3: 31, 27; Bin 4: 30, 30; Bin 5: 25, 25; Bin 6: 19, 19, 18; Bin 7: 18, 18, 18; Bin 8: 18, 17, 16; Bin 9: 16, 16, 15, 13; Bin 10: 15, 15, 14, 12; Bin 11: 12, 12, 9.

(f) An optimal packing with 10 bins exists. Bin 1: 36, 24; Bin 2: 32, 16, 12; Bin 3: 31, 17, 12; Bin 4: 30, 30; Bin 5: 27, 18, 15; Bin 6: 26, 18, 16; Bin 7: 25, 19, 16; Bin 8: 25, 19, 15; Bin 9: 18, 18, 12, 9; Bin 10: 18, 15, 14, 13.

63. (a) Packing boxes of the same height into crates: packing want ads into a newspaper page.

(b) We assume, without loss of generality, $p \geq q$. One heuristic, similar to first-fit, orders the rectangles $p \times q$ as in a dictionary (i.e., $p \times q$ listed prior to $r \times s$ if $p > r$ or $p = r$ and $q \geq s$). It then puts the rectangles in place in layers in a first-fit manner; that is, do not put a rectangle into a second layer until all positions on the first layer are filled. However, extra room in the first layer is "wasted."

(c) The problem of packing rectangles of width 1 in an $m \times 1$ rectangle is a special case of the two-dimensional problem, equivalent to the bin-packing problem we have discussed.

(d) Two 1×10 rectangles cannot be packed into a 5×4 rectangle, even though there would be an area of 20 in this rectangle.

65. There is an example of a bin-packing problem for which a given list takes a certain number of bins, and when an item is deleted from the list, more bins are required. In this example, the deleted item is not first in the list.

67. (a) Graphs (b), (d), (e), and (f) can be colored with three colors, but graphs (a) and (c) cannot.

(b) Graphs (a), (b), (d), (e), and (f) can be colored with four colors, but graph (c) cannot.

(c) The chromatic number for graphs (a) through (f) are, respectively, 4, 3, 5, 3, 2, and 2.

69. (a) Construct the graph shown below:

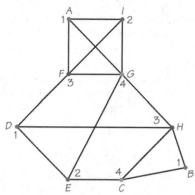

(b) The vertices of this graph can be colored with no fewer than four colors (1, 2, 3, 4 are used to denote the colors in the figure above). Hence, four tanks can be used to display the fish.

(c) The coloring in (a) shows that one can display two types of fish in three of the tanks, and three types of fish in one tank. Since 4 does not divide 9, one cannot do better.

71. (a) The graph for this situation is shown below. The vertices can be labeled with the colors 1, 2, 3 as shown.

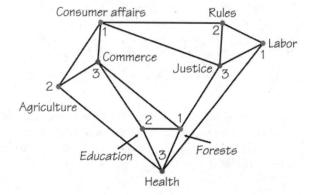

(b) Since the vertices can be colored with three colors (and no fewer), the minimum number of time slots for scheduling the committees is three.

(c) The committees can be scheduled in three rooms during each time slot. This might be significant if there were only three rooms that had microphone systems.

73. (a) To solve this problem, draw a graph by joining the vertices representing two committees if there is no x in the row and column of the table for these two committees.

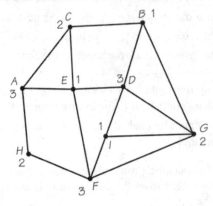

(b) The vertices of this graph can be colored with no fewer than three colors (1, 2, 3 are used to denote the colors in the figure above). Hence, three time slots are required.

(c) It is possible to three-color this graph so that each of the three colors is used three times. This means that one needs three rooms to arrange the scheduling of the nine committees.

75. Start at any vertex of the tree and label this vertex with color 1; color any vertex attached by an edge to this vertex with color 2. Continue to color the vertices in the tree in this manner, alternating the use of colors. If some vertex were attached to both a vertex colored 1 and another vertex colored 2, at some stage this would imply the graph had a circuit (of odd length), which is not possible, since trees have no circuits of any length.

77. The edge-coloring numbers for graphs (a) through (f) of Exercise 67 are, respectively, 6, 8, 6, 3, 3, and 4. The minimum edge-coloring number of any graph is either the maximal valence of any vertex in the graph or one more than the maximal valence. (This fact was discovered by the Russian mathematician Vizing.)

79. (a) Graph (a) four colors; graph (b) two colors; graph (c) four colors; graph (d) four colors; graph (e) two colors; graph (f) three colors.

(b) Coloring the maps of countries in an atlas would be one application of face colorings of graphs.

81. The minimum number of terrariums is three.

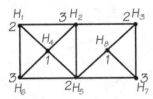

83. The minimum number of play groups is 3 because the chromatic number for this graph is 3. Since there are 7 children, the only number of play groups which could be set up which is conflict free would be if each child formed his/her own play group! With 3 play groups one can form two groups of size 2 and one group of size 3.

Chapter 4
Linear Programming

Exercise Solutions

1. (a) $2x + 3y = 12$

y-intercept: Substitute $x = 0$.

$2(0) + 3y = 12$

$0 + 3y = 12$

$3y = 12 \Rightarrow y = \frac{12}{3} = 4$

y-intercept is $(0, 4)$.

x-intercept: Substitute $y = 0$.

$2x + 3(0) = 12$

$2x + 0 = 12$

$2x = 12 \Rightarrow x = \frac{12}{2} = 6$

x-intercept is $(6, 0)$.

Graph:

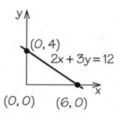

(b) $3x + 5y = 30$

y-intercept: Substitute $x = 0$.

$3(0) + 5y = 30$

$0 + 5y = 30$

$5y = 30 \Rightarrow y = \frac{30}{5} = 6$

y-intercept is $(0, 6)$.

x-intercept: Substitute $y = 0$.

$3x + 5(0) = 30$

$3x + 0 = 30$

$3x = 30 \Rightarrow x = \frac{30}{3} = 10$

x-intercept is $(10, 0)$.

Graph:

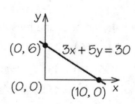

(c) $4x + 3y = 24$

y-intercept: Substitute $x = 0$.

$4(0) + 3y = 24$

$0 + 3y = 24$

$3y = 24 \Rightarrow y = \frac{24}{3} = 8$

y-intercept is $(0, 8)$.

x-intercept: Substitute $y = 0$.

$4x + 3(0) = 24$

$4x + 0 = 24$

$4x = 24 \Rightarrow x = \frac{24}{4} = 6$

x-intercept is $(6, 0)$.

Graph:

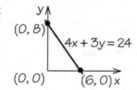

3.   **(a)** $4x + 3y = 18$ and $x = 0$

Note: This situation is shown only for the first quadrant.

The y-intercept of $4x + 3y = 18$ can be found by substituting $x = 0$.

$$4(0) + 3y = 18$$

$$0 + 3y = 18 \Rightarrow 3y = 18 \Rightarrow y = \tfrac{18}{3} = 6$$

The y-intercept is $(0,6)$.

The x-intercept of $4x + 3y = 18$ can be found by substituting $y = 0$.

$$4x + 3(0) = 18$$

$$4x + 0 = 18 \Rightarrow 4x = 18 \Rightarrow x = \tfrac{18}{4} = 4.5$$

The x-intercept is $(4.5, 0)$.

$x = 0$ represents a vertical line, namely the y-axis.

To find the point of intersection, substitute $x = 0$ into $4x + 3y = 18$.

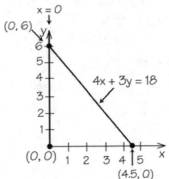

$$4(0) + 3y = 18$$

$$0 + 3y = 18$$

$$3y = 18 \Rightarrow y = \tfrac{18}{3} = 6$$

The point of intersection is therefore $(0,6)$.

(b) $5x + 3y = 45$ and $y = -5$

Note: This situation is shown only for the first and fourth quadrants.

The y-intercept of $5x + 3y = 45$ can be found by substituting $x = 0$.

$$5(0) + 3y = 45$$

$$0 + 3y = 45$$

$$3y = 45 \Rightarrow y = \tfrac{45}{3} = 15$$

The y-intercept is $(0,15)$.

The x-intercept of $5x + 3y = 45$ can be found by substituting $y = 0$.

$$5x + 3(0) = 45$$

$$5x + 0 = 45$$

$$5x = 45 \Rightarrow x = \tfrac{45}{5} = 9$$

The x-intercept is $(9,0)$.

$y = -5$ represents a horizontal line, which lies below the x-axis.

To find the point of intersection, substitute $y = -5$ into $5x + 3y = 45$.

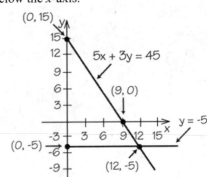

$$5x + 3(-5) = 45$$

$$5x + (-15) = 45$$

$$5x - 15 = 45$$

$$5x = 60 \Rightarrow x = \tfrac{60}{5} = 12$$

The point of intersection is therefore $(12, -5)$.

Continued on next page

3. continued

(c) $5x + 3y = 45$ and $x = 3$

Note: This situation is shown only for the first and fourth quadrants.

Graph $5x + 3y = 45$ as was done in part (b).

$x = 3$ represents a vertical line, which lies to the right of the y-axis.

To find the point of intersection, substitute $x = 3$ into $5x + 3y = 45$.

$$5(3) + 3y = 45$$
$$15 + 3y = 45$$
$$3y = 30 \Rightarrow y = \tfrac{30}{3} = 10$$

The point of intersection is therefore $(3, 10)$.

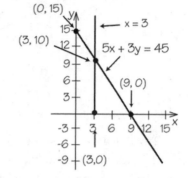

5. Note: These situations are shown only for the first quadrant.

(a) $x + y = 10$ and $x + 2y = 14$

The y-intercept of $x + y = 10$ can be found by substituting $x = 0$.

$$0 + y = 10$$
$$y = 10$$

The y-intercept is $(0, 10)$.

The x-intercept of $x + y = 10$ can be found by substituting $y = 0$.

$$x + 0 = 10$$
$$x = 10$$

The x-intercept is $(10, 0)$.

The y-intercept of $x + 2y = 14$ can be found by substituting $x = 0$.

$$0 + 2y = 14$$
$$2y = 14$$
$$y = \tfrac{14}{2} = 7$$

The y-intercept is $(0, 7)$.

The x-intercept of $x + 2y = 14$ can be found by substituting $y = 0$.

$$x + 2(0) = 14$$
$$x + 0 = 14$$
$$x = 14$$

The x-intercept is $(14, 0)$.

To find the point of intersection, we can multiply both sides of $x + y = 10$ by -1, and add the result to $x + 2y = 14$.

$$-x - y = -10$$
$$\underline{x + 2y = 14}$$
$$y = 4$$

Substitute $y = 4$ into $x + y = 10$ to solve to x.

$$x + 4 = 10 \Rightarrow x = 6$$

Continued on next page

5. (a) continued

The point of intersection is therefore $(6,4)$.

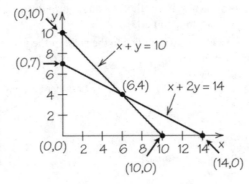

(b) $y - 2x = 0$ and $x = 2$

$x = 2$ represents a vertical line, which lies to the right of the y-axis.

By substituting either $x = 0$ or $y = 0$, we see that $y - 2x = 0$ passes through $(0,0)$, the origin. By substituting an arbitrary value (except 0) for one of the variables, we can find another point that lies on the graph of $y - 2x = 0$. Since we seek the point of intersection with the graph of $x = 2$, we could use this value.

$$y - 2(2) = 0$$
$$y - 4 = 0$$
$$y = 4$$

Thus, a second point on the graph of $y - 2x = 0$ is $(2,4)$. This is also the point of intersection between the two lines.

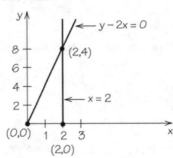

7. Note: These situations are shown only for the first quadrant.
   (a) $x \geq 3$

   The graph of $x = 3$ represents a vertical line with x-intercept $(3,0)$. Since any value of x to the right of this line is greater than 3, we shade the portion to the right of this vertical line.

   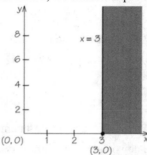

   (b) $y \geq 8$

   The graph of $y = 8$ represents a horizontal line with y-intercept $(0,8)$. Since any value of y above this line is greater than 8, we shade the portion above this horizontal line.

   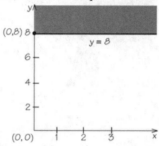

   (c) $3x + 2y \leq 18$

   The y-intercept of $3x + 2y = 18$ can be found by substituting $x = 0$.
   $$3(0) + 2y = 18 \Rightarrow 0 + 2y = 18 \Rightarrow 2y = 18 \Rightarrow y = \tfrac{18}{2} = 9$$
   The y-intercept is $(0,9)$.

   The x-intercept of $3x + 2y = 18$ can be found by substituting $y = 0$.
   $$3x + 2(0) = 18 \Rightarrow 3x + 0 = 18 \Rightarrow 3x = 18 \Rightarrow x = \tfrac{18}{3} = 6$$
   The x-intercept is $(6,0)$.

   We draw a line connecting these points. Testing the point $(0,0)$, we have the statement $3(0) + 2(0) \leq 18$ or $0 \leq 18$. This is a true statement, thus we shade the half-plane containing our test point, the down side of the line.

   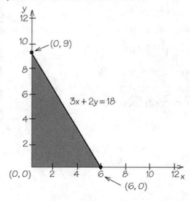

   Continued on next page

7.   continued

    (d) $7x + 2y \le 42$

The y-intercept of $7x + 2y = 42$ can be found by substituting $x = 0$.

$$7(0) + 2y = 42 \Rightarrow 0 + 2y = 42 \Rightarrow 2y = 42 \Rightarrow y = \tfrac{42}{2} = 21$$

The y-intercept is $(0, 21)$.

The x-intercept of $7x + 2y = 42$ can be found by substituting $y = 0$.

$$7x + 2(0) = 42 \Rightarrow 7x + 0 = 42 \Rightarrow 7x = 42 \Rightarrow x = \tfrac{42}{7} = 6$$

The x-intercept is $(6, 0)$.

We draw a line connecting these points. Testing the point $(0, 0)$, we have the statement $7(0) + 2(0) \le 42$ or $0 \le 42$. This is a true statement, thus we shade the half-plane containing our test point, the down side of the line.

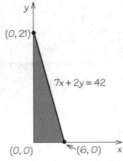

9.   **(a)** $6x + 4y \le 300$

    (b) $30x + 72y \le 420$

11. $x \ge 0;\ y \ge 0;\ 2x + y \le 10$

The constraints of $x \ge 0$ and $y \ge 0$ indicate that we are restricted to the upper right quadrant created by the x-axis and y-axis.

The y-intercept of $2x + y = 10$ can be found by substituting $x = 0$.

$$2(0) + y = 10 \Rightarrow 0 + y = 10 \Rightarrow y = 10$$

The y-intercept is $(0, 10)$.

The x-intercept of $2x + y = 10$ can be found by substituting $y = 0$.

$$2x + 0 = 10 \Rightarrow 2x = 10 \Rightarrow x = \tfrac{10}{2} = 5$$

The x-intercept is $(5, 0)$.

We draw a line connecting these points. Testing the point $(0, 0)$, we have the statement $2(0) + 0 \le 10$ or $0 \le 10$. This is a true statement, thus we shade the half-plane containing our test point, the down side of the line, which is contained in the upper right quadrant.

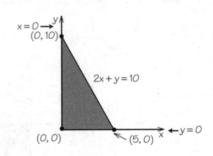

13. $x \geq 0$; $y \geq 0$; $2x + 5y \leq 60$

The constraints of $x \geq 0$ and $y \geq 0$ indicate that we are restricted to the upper right quadrant created by the x-axis and y-axis.

The y-intercept of $2x + 5y = 60$ can be found by substituting $x = 0$.

$$2(0) + 5y = 60 \Rightarrow 0 + 5y = 60 \Rightarrow y = \frac{60}{5} = 12$$

The y-intercept is $(0, 12)$.

The x-intercept of $2x + 5y = 60$ can be found by substituting $y = 0$.

$$2x + 5(0) = 60 \Rightarrow 2x + 0 = 60 \Rightarrow 2x = 60 \Rightarrow x = \frac{60}{2} = 30$$

The x-intercept is $(30, 0)$.

We draw a line connecting these points. Testing the point $(0, 0)$, we have the statement $2(0) + 5(0) \leq 60$ or $0 \leq 60$. This is a true statement, thus we shade the half-plane containing our test point, the down side of the line, which is contained in the upper right quadrant.

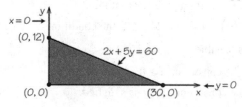

15. $x \geq 0$; $y \geq 4$; $x + y \leq 20$

The constraints of $x \geq 0$ and $y \geq 4$ indicate that we are restricted to the upper right quadrant, above the horizontal line $y = 4$.

The point of intersection between $y = 4$ and $x + y = 20$ can be found by substituting $y = 4$ into $x + y = 20$.

$$x + 4 = 20 \Rightarrow x = 16$$

Thus, the point of intersection is $(16, 4)$.

The y-intercept of $x + y = 20$ can be found by substituting $x = 0$.

$$0 + y = 20 \Rightarrow y = 20$$

The y-intercept is $(0, 20)$.

The x-intercept of $x + y = 20$ can be found by substituting $y = 0$.

$$x + 0 = 20 \Rightarrow x = 20$$

The x-intercept is $(20, 0)$.

We draw a line connecting these points. Testing the point $(0, 0)$, we have the statement $0 + 0 \leq 20$ or $0 \leq 20$. This is a true statement, thus we shade the half-plane containing our test point, the down side of the line, which is contained in the upper right quadrant above the horizontal line $y = 4$.

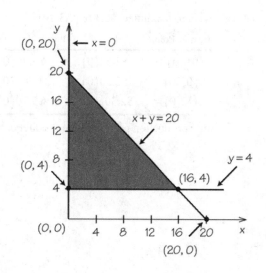

17. For Exercise 11: $x \geq 0$; $y \geq 0$; $2x + y \leq 10$

$(2,4)$:   Since $2 \geq 0$, the constraint $x \geq 0$ is satisfied.

Since $4 \geq 0$, the constraint $y \geq 0$ is satisfied.

Since $2(2) + 4 = 4 + 4 = 8 \leq 10$, the condition $2x + y \leq 10$ is satisfied.

Thus, $(2,4)$ is feasible.

$(10,6)$:   Since $10 \geq 0$, the constraint $x \geq 0$ is satisfied.

Since $6 \geq 0$, the constraint $y \geq 0$ is satisfied.

Since $2(10) + 6 = 20 + 6 = 26 > 10$, the condition $2x + y \leq 10$ is not satisfied.

Thus, $(10,6)$ is not feasible.

For Exercise 13: $x \geq 0$; $y \geq 0$; $2x + 5y \leq 60$

$(2,4)$:   Since $2 \geq 0$, the constraint $x \geq 0$ is satisfied.

Since $4 \geq 0$, the constraint $y \geq 0$ is satisfied.

Since $2(2) + 5(4) = 4 + 20 = 24 \leq 60$, the condition $x + 2y \leq 12$ is satisfied.

Thus, $(2,4)$ is feasible.

$(10,6)$:   Since $10 \geq 0$, the constraint $x \geq 0$ is satisfied.

Since $6 \geq 0$, the constraint $y \geq 0$ is satisfied.

Since $2(10) + 5(6) = 20 + 30 = 50 \leq 60$, the condition $x + 2y \leq 12$ is satisfied.

Thus, $(10,6)$ is feasible.

For Exercise 15: $x \geq 0$; $y \geq 4$; $x + y \leq 20$

$(2,4)$:   Since $2 \geq 0$, the constraint $x \geq 0$ is satisfied.

Since $4 \geq 4$, the constraint $y \geq 4$ is satisfied.

Since $2 + 4 = 6 \leq 20$, the condition $x + y \leq 20$ is satisfied.

Thus, $(2,4)$ is feasible. Note: It is on the boundary.

$(10,6)$:   Since $10 \geq 0$, the constraint $x \geq 0$ is satisfied.

Since $6 \geq 4$, the constraint $y \geq 4$ is satisfied.

Since $10 + 6 = 16 \leq 20$, the condition $x + y \leq 20$ is satisfied.

Thus, $(10,6)$ is feasible.

19. We wish to maximize $\$2.30x + \$3.70y$.

Corner Point	Value of the Profit Formula: $\$2.30x + \$3.70y$						
$(0,0)$	$\$2.30(0)$	+	$\$3.70(0)$	=	$\$0.00$ +	$\$0.00$ =	$\$0.00$
$(0,30)$	$\$2.30(0)$	+	$\$3.70(30)$	=	$\$0.00$ +	$\$111.00$ =	$\$111.00*$
$(12,0)$	$\$2.30(12)$	+	$\$3.70(0)$	=	$\$27.60$ +	$\$0.00$ =	$\$27.60$

Optimal production policy: Make 0 skateboards and 30 dolls for a profit of $111.

21. Note: These situations are shown only for the first quadrant.

    (a) $5x + 4y = 22$ and $5x + 10y = 40$

The y-intercept of $5x + 4y = 22$ can be found by substituting $x = 0$.

$$5(0) + 4y = 22$$
$$0 + 4y = 22$$
$$4y = 22$$
$$y = \tfrac{22}{4} = 5.5$$

The y-intercept is $(0, 5.5)$.

The x-intercept of $5x + 4y = 22$ can be found by substituting $y = 0$.

$$5x + 4(0) = 22$$
$$5x + 0 = 22$$
$$5x = 22$$
$$x = \tfrac{22}{5} = 4.4$$

The x-intercept is $(4.4, 0)$.

The y-intercept of $5x + 10y = 40$ can be found by substituting $x = 0$.

$$5(0) + 10y = 40$$
$$0 + 10y = 40$$
$$y = \tfrac{40}{10} = 4$$

The y-intercept is $(0, 4)$.

The x-intercept of $5x + 10y = 40$ can be found by substituting $y = 0$.

$$5x + 10(0) = 40$$
$$5x + 0 = 40$$
$$5x = 40$$
$$x = \tfrac{40}{4} = 8$$

The x-intercept is $(8, 0)$.

To find the point of intersection, we can multiply both sides of $5x + 4y = 22$ by -1, and adding the result to $5x + 10y = 40$.

$$-5x - 4y = -22$$
$$\underline{5x + 10y = 40}$$
$$6y = 18 \Rightarrow y = \tfrac{18}{6} = 3$$

Substitute $y = 3$ into $5x + 10y = 40$ and solve for x.

$$5x + 10(3) = 40 \Rightarrow 5x + 30 = 40 \Rightarrow 5x = 10 \Rightarrow x = \tfrac{10}{5} = 2$$

The point of intersection is therefore $(2, 3)$.

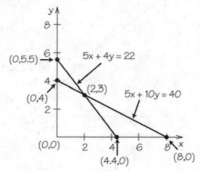

Continued on next page

21. continued

    (b) $x + y = 7$ and $3x + 4y = 24$

        The y-intercept of $x + y = 7$ can be found by substituting $x = 0$.

$$0 + y = 7$$
$$y = 7$$

        The y-intercept is $(0, 7)$.

        The x-intercept of $x + y = 7$ can be found by substituting $y = 0$.

$$x + 0 = 7$$
$$x = 7$$

        The x-intercept is $(7, 0)$.

        The y-intercept of $3x + 4y = 24$ can be found by substituting $x = 0$.

$$3(0) + 4y = 24$$
$$0 + 4y = 24$$
$$y = \tfrac{24}{4} = 6$$

        The y-intercept is $(0, 6)$.

        The x-intercept of $3x + 4y = 24$ can be found by substituting $y = 0$.

$$3x + 4(0) = 24$$
$$3x + 0 = 24$$
$$3x = 24$$
$$x = \tfrac{24}{3} = 8$$

        The x-intercept is $(8, 0)$.

        To find the point of intersection, we can multiply both sides of $x + y = 7$ by -3, and adding the result to $3x + 4y = 24$

$$-3x - 3y = -21$$
$$\underline{3x + 4y = 24}$$
$$y = 3$$

        Substitute $y = 3$ into $3x + 4y = 24$ and solve for x.

$$3x + 4(3) = 24 \Rightarrow 3x + 12 = 24 \Rightarrow 3x = 12 \Rightarrow x = \tfrac{12}{3} = 4$$

        The point of intersection is therefore $(4, 3)$.

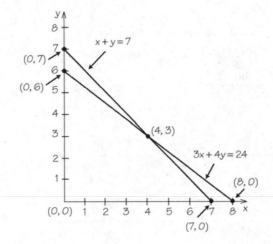

23. $x \geq 0;\, y \geq 0;\, 2x + y \leq 4;\, 4x + 4y \leq 12$

The constraints of $x \geq 0$ and $y \geq 0$ indicate that we are restricted to the upper right quadrant created by the x-axis and y-axis.

The y-intercept of $2x + y = 4$ can be found by substituting $x = 0$.

$$2(0) + y = 4 \Rightarrow 0 + y = 4 \Rightarrow y = 4$$

The y-intercept is $(0, 4)$.

The x-intercept of $2x + y = 4$ can be found by substituting $y = 0$.

$$2x + 0 = 4 \Rightarrow 2x = 4 \Rightarrow x = \tfrac{4}{2} = 2$$

The x-intercept is $(2, 0)$.

We draw a line connecting these points. Testing the point $(0,0)$, we have the statement $2(0) + 0 \leq 4$ or $0 \leq 4$. This is a true statement.

The y-intercept of $4x + 4y = 12$ can be found by substituting $x = 0$.

$$4(0) + 4y = 12 \Rightarrow 0 + 4y = 12 \Rightarrow y = \tfrac{12}{4} = 3$$

The y-intercept is $(0, 3)$.

The x-intercept of $4x + 4y = 12$ can be found by substituting $y = 0$.

$$4x + 4(0) = 12 \Rightarrow 4x + 0 = 12 \Rightarrow 4x = 12 \Rightarrow x = \tfrac{12}{4} = 3$$

The x-intercept is $(3, 0)$.

We draw a line connecting these points. Testing the point $(0,0)$, we have the statement $4(0) + 4(0) \leq 12$ or $0 \leq 12$. This is a true statement.

Thus, we shade the part of the plane in the upper right quadrant which is on the down side of both the lines $2x + y = 4$ and $4x + 4y = 12$.

Three of the corner points, $(0,0)$, $(0,3)$, and $(2,0)$ lie on the coordinate axes. The fourth corner point is the point of intersection between the lines $2x + y = 4$ and $4x + 4y = 12$. We can find this by multiplying both sides of $2x + y = 4$ by -4, and adding the result to $4x + 4y = 12$.

$$-8x - 4y = -16$$
$$\underline{4x + 4y = 12}$$
$$-4x \qquad = -4 \Rightarrow x = \tfrac{-4}{-4} = 1$$

Substitute $x = 1$ into $2x + y = 4$ and solve for y.

$$2(1) + y = 4 \Rightarrow 2 + y = 4 \Rightarrow y = 2$$

The point of intersection is therefore $(1, 2)$.

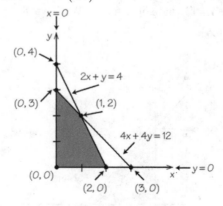

25. $x \geq 4$; $y \geq 0$; $5x + 4y \leq 60$; $x + y \leq 13$

The constraints of $x \geq 4$ and $y \geq 0$ indicate that we are restricted to the upper right quadrant, to the right of the vertical line $x = 4$.

The y-intercept of $5x + 4y = 60$ can be found by substituting $x = 0$.

$$5(0) + 4y = 60 \Rightarrow 0 + 4y = 60 \Rightarrow y = \tfrac{60}{4} = 15$$

The y-intercept is $(0, 15)$.

The x-intercept of $5x + 4y = 60$ can be found by substituting $y = 0$.

$$5x + 4(0) = 60 \Rightarrow 5x + 0 = 60 \Rightarrow 5x = 60 \Rightarrow x = \tfrac{60}{5} = 12$$

The x-intercept is $(12, 0)$.

We draw a line connecting these points. Testing the point $(0, 0)$, we have the statement $5(0) + 4(0) \leq 60$ or $0 \leq 60$. This is a true statement.

The y-intercept of $x + y = 13$ can be found by substituting $x = 0$.

$$0 + y = 13 \Rightarrow y = 13$$

The y-intercept is $(0, 13)$.

The x-intercept of $x + y = 13$ can be found by substituting $y = 0$.

$$x + 0 = 13 \Rightarrow x = 13$$

The x-intercept is $(13, 0)$.

We draw a line connecting these points. Testing the point $(0, 0)$, we have the statement $0 + 0 \leq 13$ or $0 \leq 13$. This is a true statement.

Thus, we shade the part of the plane in the upper right quadrant which is on the down side of both the lines $5x + 4y = 60$ and $x + y = 13$, which is also to the right of the vertical line $x = 4$.

Two of the corner points, $(4, 0)$ and $(12, 0)$, lie on the coordinate axes. The third corner point is the point of intersection between the lines $x + y = 13$ and $x = 4$. We can find this by substituting $x = 4$ into $x + y = 13$ to solve to y. We have, $4 + y = 13 \Rightarrow y = 9$. The point of intersection is therefore $(4, 9)$.

The fourth corner point is the point of intersection between the lines $5x + 4y = 60$ and $x + y = 13$. We can find this by multiplying both sides of $x + y = 13$ by -4, and adding the result to $5x + 4y = 60$.

$$
\begin{array}{r}
-4x - 4y = -52 \\
5x + 4y = 60 \\
\hline
x \qquad\;\; = 8
\end{array}
$$

Substitute $x = 8$ into $x + y = 13$ and solve for y. We have $8 + y = 13 \Rightarrow y = 5$. The point of intersection is therefore $(8, 5)$.

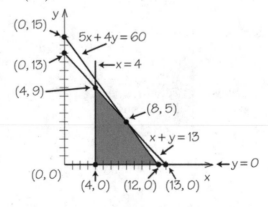

27. Maximize $P = 3x + 2y$ subject to $x \geq 3;\ y \geq 2;\ x + y \leq 10;\ 2x + 3y \leq 24$

We need to first graph the feasible region while finding the corner points.

The constraints of $x \geq 3$ and $y \geq 2$ indicate that we are restricted to the upper right quadrant, to the right of the vertical line $x = 3$ and above the horizontal line $y = 2$.

The point of intersection between $x = 3$ and $y = 2$ is $(3,2)$.

The y-intercept of $x + y = 10$ can be found by substituting $x = 0$.

$$0 + y = 10 \Rightarrow y = 10$$

The y-intercept is $(0,10)$.

The x-intercept of $x + y = 10$ can be found by substituting $y = 0$.

$$x + 0 = 10 \Rightarrow x = 10$$

The x-intercept is $(10,0)$.

The y-intercept of $2x + 3y = 24$ can be found by substituting $x = 0$.

$$2(0) + 3y = 24 \Rightarrow 0 + 3y = 24 \Rightarrow 3y = 24 \Rightarrow y = \tfrac{24}{3} = 8$$

The y-intercept is $(0,8)$.

The x-intercept of $2x + 3y = 24$ can be found by substituting $y = 0$.

$$2x + 3(0) = 24 \Rightarrow 2x + 0 = 24 \Rightarrow 2x = 24 \Rightarrow x = \tfrac{24}{2} = 12$$

The x-intercept is $(12,0)$.

Continued on next page

27. continued

Testing the point $(0,0)$ in both $x+y \leq 10$ and $2x+3y \leq 24$, we have the following.

$$0+0 = 0 \leq 10 \text{ and } 2(0)+3(0) = 0 \leq 24$$

Since these are both true statements, we shade the down side of the line of both lines which is contained in the upper right quadrant to the right of the vertical line $x = 3$ and above the horizontal line $y = 2$.

The point of intersection between $x = 3$ and $2x+3y = 24$ can be found by substituting $x = 3$ into $2x+3y = 24$. We have $2(3)+3y = 24 \Rightarrow 6+3y = 24 \Rightarrow 3y = 18 \Rightarrow y = \frac{18}{3} = 6$. Thus, the point of intersection is $(3,6)$.

The point of intersection between $y = 2$ and $x+y = 10$ can be found by substituting $y = 2$ into $x+y = 10$. We have, $x+2 = 10 \Rightarrow x = 8$. Thus, the point of intersection is $(8,2)$.

The final corner point is the point of intersection between $x+y = 10$ and $2x+3y = 24$. We can find this by multiplying both sides of $x+y = 10$ by -2, and adding the result to $2x+3y = 24$.

$$-2x-2y = -20$$
$$\underline{2x+3y = 24}$$
$$y = 4$$

Substitute $y = 4$ into $x+y = 10$ and solve for x. We have, $x+4 = 10 \Rightarrow x = 6$. The point of intersection is therefore $(6,4)$.

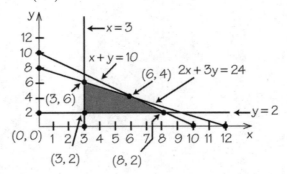

We wish to maximize $P = 3x+2y$.

Corner Point	Value of the Profit Formula: $3x+2y$
$(3,6)$	$3(3)$ + $2(6)$ = 9 + 12 = 21
$(3,2)$	$3(3)$ + $2(2)$ = 9 + 4 = 13
$(8,2)$	$3(8)$ + $2(2)$ = 24 + 4 = 28*
$(6,4)$	$3(6)$ + $2(4)$ = 18 + 8 = 26

The maximum value occurs at the corner point $(8,2)$, where P is equal to 28.

29. Maximize $P = 5x + 2y$ subject to $x \ge 2$; $y \ge 4$; $x + y \le 10$

We need to first graph the feasible region while finding the corner points.

The constraints of $x \ge 2$ and $y \ge 4$ indicate that we are restricted to the upper right quadrant, to the right of the vertical line $x = 2$ and above the horizontal line $y = 4$.

The point of intersection between $x = 2$ and $y = 4$ is $(2, 4)$.

The y-intercept of $x + y = 10$ can be found by substituting $x = 0$.

$$0 + y = 10 \Rightarrow y = 10$$

The y-intercept is $(0, 10)$.

The x-intercept of $x + y = 10$ can be found by substituting $y = 0$.

$$x + 0 = 10 \Rightarrow x = 10$$

The x-intercept is $(10, 0)$.

Testing the point $(0, 0)$ in $x + y \le 10$, we have the following.

$$0 + 0 = 0 \le 10$$

Since this is a true statement, we shade the down side of this line which is contained in the upper right quadrant to the right of the vertical line $x = 2$ and above the horizontal line $y = 4$.

The point of intersection between $x = 2$ and $x + y = 10$ can be found by substituting $x = 2$ into $x + y = 10$. We have $2 + y = 10 \Rightarrow y = 8$. Thus, the point of intersection is $(2, 8)$.

The point of intersection between $y = 4$ and $x + y = 10$ can be found by substituting $y = 4$ into $x + y = 10$. We have $x + 4 = 10 \Rightarrow x = 6$. Thus, the point of intersection is $(6, 4)$.

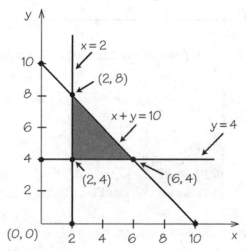

We wish to maximize $P = 5x + 2y$.

Corner Point	Value of the Profit Formula: $5x + 2y$
$(2, 4)$	$5(2) \; + \; 2(4) \; = \; 10 \; + \; 8 \; = \; 18$
$(2, 8)$	$5(2) \; + \; 2(8) \; = \; 10 \; + \; 16 \; = \; 26$
$(6, 4)$	$5(6) \; + \; 2(4) \; = \; 30 \; + \; 8 \; = \; 38^*$

The maximum value occurs at the corner point $(6, 4)$, where P is equal to 38.

31. (a) The optimal corner point for Exercise 30 being $\left(\frac{13}{7},0\right)$, the coordinates of $Q = (2,0)$.

(b) $(2,0)$ is not a feasible point.

(c) The profit at $(2,0)$ is greater than the profit at $\left(\frac{13}{7},0\right)$.

(d) Since $0 \geq 0$ and $3 \geq 0$ the constraints $x \geq 0$ and $y \geq 0$ are satisfied, respectively. Also, since $7(0) + 4(3) = 0 + 12 = 12 \leq 13$, the constraint $7x + 4y \leq 13$ is satisfied. Thus, $R = (0,3)$ is feasible. The profit at R is $21(0) + 11(3) = 33$. This is less than the profit at Q.

(e) Solving maximization problems involving linear constraints but where the variables are required to take on integer values cannot be solved by first solving the associated linear programming problem and rounding the answer to the nearest integers. This example shows the rounded value used to obtain an integer solution may not be feasible. Even if the rounded value is feasible it may not be optimal. "Integer programming" is unfortunately a much harder problem to solve than linear programming.

For Exercises 33 – 43, part (e) (using a simplex algorithm program) will not be addressed in the solutions.

33. (a) Let x be the number of oil changes and y be the number tune-ups.

	Time (8,000 min)	Minimums	Profit
Oil changes, x	20	0	$15
Tune-ups, y	100	0	$65

(b) Profit formula: $P = \$15x + \$65y$

Constraints: $x \geq 0$ and $y \geq 0$ (minimums); $20x + 100y \leq 8,000$ (time)

(c) Feasible region:

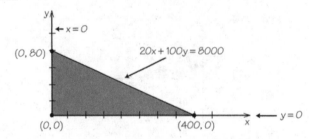

Corner points: The corner points are intercepts on the axes. These are $(0,0)$, $(0,80)$, and $(400,0)$.

Continued on next page

33. continued

(d) We wish to maximize $\$15x + \$65y$.

Corner Point	Value of the Profit Formula: $\$15x + \$65y$								
$(0,0)$	$\$15(0)$	$+$	$\$65(0)$	$=$	$\$0$	$+$	$\$0$	$=$	$\$0$
$(0,80)$	$\$15(0)$	$+$	$\$65(80)$	$=$	$\$0$	$+$	$\$5200$	$=$	$\$5200$
$(400,0)$	$\$15(400)$	$+$	$\$65(0)$	$=$	$\$6000$	$+$	$\$0$	$=$	$\$6000*$

Optimal production policy: Schedule 400 oil changes and no tune-ups.

With non-zero minimums, the constraints are as follows.

$$x \geq 50 \text{ and } y \geq 20 \text{ (minimums)}; 20x + 100y \leq 8,000 \text{ (time)}$$

The feasible region looks like the following.

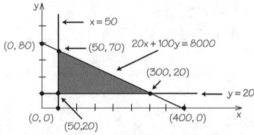

Corner points: One corner point is the point of intersection between $x = 50$ and $y = 20$, namely $(50, 20)$. Another is the point of intersection between $x = 50$ and $20x + 100y = 8000$. Substituting $x = 50$ into $20x + 100y = 8000$, we have the following.

$$20(50) + 100y = 8000 \Rightarrow 1000 + 100y = 8000 \Rightarrow 100y = 7000 \Rightarrow y = 70$$

Thus, $(50, 70)$ is the second corner point. The third corner point is the point of intersection between $y = 20$ and $20x + 100y = 8000$. Substituting $y = 20$ into $20x + 100y = 8000$, we have the following.

$$20x + 100(20) = 8000 \Rightarrow 20x + 2000 = 8000 \Rightarrow 20x = 6000 \Rightarrow x = 300$$

Thus, $(300, 20)$ is the third corner point.

We wish to maximize $\$15x + \$65y$.

Corner Point	Value of the Profit Formula: $\$15x + \$65y$								
$(50, 20)$	$\$15(50)$	$+$	$\$65(20)$	$=$	$\$750$	$+$	$\$1300$	$=$	$\$2050$
$(50, 70)$	$\$15(50)$	$+$	$\$65(70)$	$=$	$\$750$	$+$	$\$4550$	$=$	$\$5300$
$(300, 20)$	$\$15(300)$	$+$	$\$65(20)$	$=$	$\$4500$	$+$	$\$1300$	$=$	$\$5800*$

Optimal production policy: Schedule 300 oil changes and 20 tune-ups.

35. (a) Let x be the number of routine visits and y be the number of comprehensive visits.

	Doctor Time (1800 min)	Minimums	Profit
Routine, x visits	5	0	$30
Comprehensive, y visits	25	0	$50

(b) Profit formula: $P = \$30x + \$50y$

Constraints: $x \geq 0$ and $y \geq 0$ (minimums); $5x + 25y \leq 1800$ (time)

(c) Feasible region:

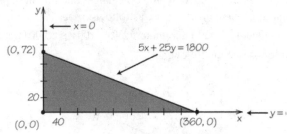

Corner points: The corner points are intercepts on the axes. These are $(0,0)$, $(0,72)$, and $(360,0)$.

(d) We wish to maximize $\$30x + \$50y$.

Corner Point	Value of the Profit Formula: $\$30x + \$50y$						
$(0,0)$	$\$30(0)$	+	$\$50(0)$	=	$\$0$ +	$\$0$ =	$\$0$
$(0,72)$	$\$30(0)$	+	$\$50(72)$	=	$\$0$ +	$\$3600$ =	$\$3600$
$(360,0)$	$\$30(360)$	+	$\$50(0)$	=	$\$10,800$ +	$\$0$ =	$\$10,800\ast$

Optimal production policy: Schedule 360 routine visits and no comprehensive visits. With non-zero minimums, the constraints are as follows.

$$x \geq 20 \text{ and } y \geq 30 \text{ (minimums)}; 5x + 25y \leq 1800 \text{ (time)}$$

The feasible region looks like the following.

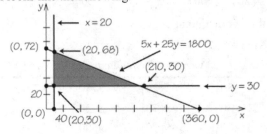

Corner points: One corner point is the point of intersection between $x = 20$ and $y = 30$, namely $(20,30)$. Another is the point of intersection between $x = 20$ and $5x + 25y = 1800$. Substituting $x = 20$ into $5x + 25y = 1800$, we have the following.

$$5(20) + 25y = 1800 \Rightarrow 100 + 25y = 1800 \Rightarrow 25y = 1700 \Rightarrow y = 68$$

Thus, $(20,68)$ is the second corner point. The third corner point is the point of intersection between $y = 30$ and $5x + 25y = 1800$. Substituting $y = 30$ into $5x + 25y = 1800$, we have the following.

$$5x + 25(30) = 1800 \Rightarrow 5x + 750 = 1800 \Rightarrow 5x = 1050 \Rightarrow x = 210$$

Continued on next page

35. continued

Thus, $(210,30)$ is the third corner point.

We wish to maximize $\$30x + \$50y$.

Corner Point	Value of the Profit Formula: $\$30x + \$50y$
$(20,30)$	$\$30(20)$ + $\$50(30)$ = $\$600$ + $\$1500$ = $\$2100$
$(20,68)$	$\$30(20)$ + $\$50(68)$ = $\$600$ + $\$3400$ = $\$4000$
$(210,30)$	$\$30(210)$ + $\$50(30)$ = $\$6300$ + $\$1500$ = $\$7800*$

Optimal production policy: Schedule 210 routine visits and 30 comprehensive visits.

37. (a) Let x be the number of hours spent on math courses and y be the number of hours spent on other courses.

	Time (48 hr)	Minimums	Value Points
Math, x courses	12	0	2
Other, y courses	8	0	1

(b) Value Point formula: $V = 2x + y$

Constraints: $x \geq 0$ and $y \geq 0$ (minimums); $12x + 8y \leq 48$ (time)

(c) Feasible region:

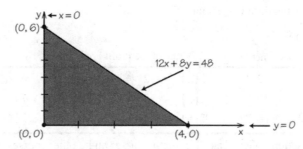

Corner points: The corner points are intercepts on the axes. These are $(0,0)$, $(0,6)$, and $(4,0)$.

(d) We wish to maximize $2x + y$.

Corner Point	Value of the Value Point Formula: $2x + y$
$(0,0)$	$2(0)$ + 0 = 0 + 0 = 0
$(0,6)$	$2(0)$ + 6 = 0 + 6 = 6
$(4,0)$	$2(4)$ + 0 = 8 + 0 = 8*

Optimal production policy: Take four math courses and no other courses.

Continued on next page

37. continued

With non-zero minimums, the constraints are $x \ge 2$ and $y \ge 2$ $\left(\text{minimums}\right); 12x + 8y \le 48$ $\left(\text{time}\right)$.
The feasible region looks like the following.

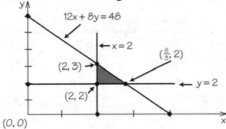

Corner points: One corner point is the point of intersection between $x = 2$ and $y = 2$, namely $\left(2, 2\right)$. Another is the point of intersection between $x = 2$ and $12x + 8y = 48$. Substituting $x = 2$ into $12x + 8y = 48$, we have the following.

$$12\left(2\right) + 8y = 48 \Rightarrow 24 + 8y = 48 \Rightarrow 8y = 24 \Rightarrow y = 3$$

Thus, $\left(2, 3\right)$ is the second corner point.

The third corner point is the point of intersection between $y = 2$ and $12x + 8y = 48$. Substituting $y = 2$ into $12x + 8y = 48$, we have the following.

$$12x + 8\left(2\right) = 48 \Rightarrow 12x + 16 = 48 \Rightarrow 12x = 32 \Rightarrow x = \frac{32}{12} = \frac{8}{3} = 2\frac{2}{3}$$

Thus, $\left(\frac{8}{3}, 2\right)$ is the third corner point.

We wish to maximize $2x + y$.

Corner Point	Value of the Profit Formula: $2x + y$							
$\left(2, 2\right)$	$2\left(2\right)$	+	2	=	4	+	2	= 6
$\left(2, 3\right)$	$2\left(2\right)$	+	3	=	4	+	3	= 7
$\left(\frac{8}{3}, 2\right)$	$2\left(\frac{8}{3}\right)$	+	2	=	$5\frac{1}{3}$	+	2	= $7\frac{1}{3}$*

Optimal production policy: Take $2\frac{2}{3}$ math courses and 2 other courses. However, one cannot take a fractional part of a course. So given the constraint on study time, the student should take two math courses and 2 other courses.

39. (a) Let x be the number of Grade A batches and y be the number of Grade B batches.

	Scrap cloth (100 lb)	Scrap Paper (120 lb)	Minimums	Profit
Grade A, x batches	25	10	0	$500
Grade B, y batches	10	20	0	$250

(b) Profit formula: $P = \$500x + \$250y$

Constraints: $x \geq 0$ and $y \geq 0$ (minimums); $25x + 10y \leq 100$ (cloth); $10x + 20y \leq 120$ (paper)

(c) Feasible region:

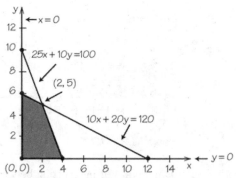

Corner points: Three of the corner points are intercepts on the axes. These are $(0,0)$, $(0,6)$, and $(4,0)$. The final corner point is the point of intersection between $25x + 10y = 100$ and $10x + 20y = 120$. We can find this by multiplying both sides of $25x + 10y = 100$ by -2, and adding the result to $10x + 20y = 120$.

$$-50x - 20y = -200$$
$$\underline{10x + 20y = 120}$$
$$-40x \qquad = -80 \Rightarrow x = \tfrac{-80}{-40} = 2$$

Substitute $x = 2$ into $25x + 10y = 100$ and solve for y. We have the following.

$$25(2) + 10y = 100 \Rightarrow 50 + 10y = 100 \Rightarrow 10y = 50 \Rightarrow y = 5$$

The point of intersection is therefore $(2,5)$.

(d) We wish to maximize $\$500x + \$250y$.

Corner Point	Value of the Profit Formula: $\$500x + \$250y$						
$(0,0)$	$\$500(0)$	+	$\$250(0)$	=	$\$0$ +	$\$0$ =	$\$0$
$(0,6)$	$\$500(0)$	+	$\$250(6)$	=	$\$0$ +	$\$1500$ =	$\$1500$
$(4,0)$	$\$500(4)$	+	$\$250(0)$	=	$\$2000$ +	$\$0$ =	$\$2000$
$(2,5)$	$\$500(2)$	+	$\$250(5)$	=	$\$1000$ +	$\$1250$ =	$\$2250*$

Optimal production policy: Make 2 grade A and 5 grade B batches.

With non-zero minimums $x \geq 1$ and $y \geq 1$, there will be no change because the optimal production policy already obeys these non-zero minimums.

41. (a) Let x be the number of cartons of regular soda and y be the number of cartons of diet soda.

	Cartons (5000)	Money ($5400)	Minimums	Profit
Regular, x cartons	1	$1.00	600	$0.10
Diet, y cartons	1	$1.20	1000	$0.11

(b) Profit formula: $P = \$0.10x + \$0.11y$

Constraints: $x \geq 600$ and $y \geq 1000$ (minimums)

$$x + y \leq 5000 \, (\text{cartons}); 1.00x + 1.20y \leq 5400 \, (\text{money})$$

(c) Feasible region:

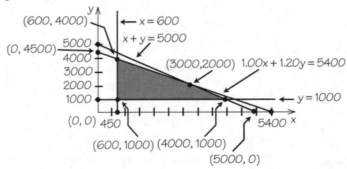

Corner points: One corner point is the point of intersection between $x = 600$ and $y = 1000$, namely $(600, 1000)$.   Another   is   the   point   of   intersection   between $y = 1000$ and $x + y = 5000$.   Substituting $y = 1000$ into $x + y = 5000$,   we   have $x + 1000 = 5000 \Rightarrow x = 4000$.   Thus, $(4000, 1000)$ is the second corner point.   The third corner   point   is   the   point   of   intersection   between $x = 600$ and $1.00x + 1.20y = 5400$. Substituting $x = 600$ into $1.00x + 1.20y = 5400$, we have the following.

$$1.00(600) + 1.20y = 5400 \Rightarrow 600 + 1.20y = 5400 \Rightarrow 1.20y = 4800 \Rightarrow y = 4000$$

Thus, $(600, 4000)$ is the third corner point.   The fourth corner point is the point of intersection   between $x + y = 5000$ and $1.00x + 1.20y = 5400$.   We can find this by multiplying both sides of $x + y = 5000$ by -1, and adding the result to $1.00x + 1.20y = 5400$.

$$-x - \quad y = -5000$$
$$\underline{1.00x + 1.20y = 5400}$$
$$0.20y = 400 \Rightarrow y = 2000$$

Substitute $y = 2000$ into $x + y = 5000$ and solve for y.   We have $x + 2000 = 5000$ or $x = 3000$.   The point of intersection is therefore $(3000, 2000)$.

(d) We wish to maximize $\$0.10x + \$0.11y$.

Corner Point	Value of the Profit Formula: $\$0.10x + \$0.11y$						
$(600, 1000)$	$\$0.10(600)$	+	$\$0.11(1000)$	=	$\$60$	+ $\$110$	= $\$170$
$(4000, 1000)$	$\$0.10(4000)$	+	$\$0.11(1000)$	=	$\$400$	+ $\$110$	= $\$510$
$(600, 4000)$	$\$0.10(600)$	+	$\$0.11(4000)$	=	$\$60$	+ $\$440$	= $\$500$
$(3000, 2000)$	$\$0.10(3000)$	+	$\$0.11(2000)$	=	$\$300$	+ $\$220$	= $\$520*$

Optimal production policy: Make 3000 cartons of regular and 2000 cartons of diet.

With   zero   minimums   there   is   no   change   in   the   optimal   production   policy   because the corresponding corner point does not touch either line from a minimum constraint.

43. (a) Let x be the number of desk lamps and y be the number of floor lamps.

	Labor (1200 hr)	Money ($4200)	Minimums	Profit
Desk, x lamps	0.8	$4	0	$2.65
Floor, y lamps	1.0	$3	0	$4.67

(b) Profit formula: $P = \$2.65x + \$4.67y$.

Constraints: $x \geq 0$ and $y \geq 0$ (minimums)

$$0.8x + 1.0y \leq 1200 \text{ (labor)}$$

$$4x + 3y \leq 4200 \text{ (money)}$$

(c) Feasible region:

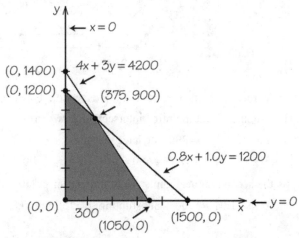

Corner points: Three of the corner points are intercepts on the axes. These are $(0,0)$, $(0,1200)$, and $(1050,0)$. The final corner point is the point of intersection between $0.8x + 1.0y = 1200$ and $4x + 3y = 4200$. We can find this by multiplying both sides of $0.8x + 1.0y = 1200$ by -3, and adding the result to $4x + 3y = 4200$.

$$-2.4x - 3y = -3600$$
$$\underline{4x + 3y = 4200}$$
$$1.6x \qquad = 600 \Rightarrow x = 375$$

Substitute $x = 375$ into $0.8x + 1.0y = 1200$ and solve for y. We have the following.

$$0.8(375) + y = 1200 \Rightarrow 300 + y = 1200 \Rightarrow y = 900$$

The point of intersection is therefore $(375, 900)$.

(d) We wish to maximize $\$2.65x + \$4.67y$.

Corner Point	Value of the Profit Formula: $\$2.65x + \$4.67y$							
$(0,0)$	$2.65(0)$	+	$4.67(0)$	=	$0.00	+	$0.00	= $0.00
$(0,1200)$	$2.65(0)$	+	$4.67(1200)$	=	$0.00	+	$5604.00	= $5604.00*
$(1050,0)$	$2.65(1050)$	+	$4.67(0)$	=	$2782.50	+	$0.00	= $2782.50
$(375,900)$	$2.65(375)$	+	$4.67(900)$	=	$993.75	+	$4203.00	= $5196.75

Optimal production policy: Make no desk lamps and 1200 floor lamps.
Continued on next page

43. continued

With non-zero minimums, the constraints are as follows.

$x \geq 150$ and $y \geq 200$ (minimums); $0.8x + 1.0y \leq 1200$ (labor); $4x + 3y \leq 4200$ (money)

The feasible region looks like the following.

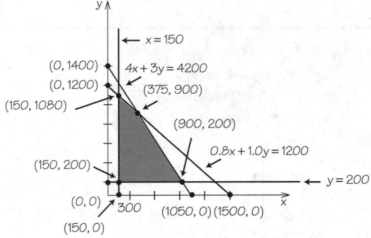

Corner points: One corner point is the point of intersection between $x = 150$ and $y = 200$, namely $(150, 200)$.   Another is the point of intersection between $y = 200$ and $4x + 3y = 4200$. Substituting $y = 200$ into $4x + 3y = 4200$, we have the following.

$$4x + 3(200) = 4200 \Rightarrow 4x + 600 = 4200 \Rightarrow 4x = 3600 \Rightarrow x = 900$$

Thus, $(900, 200)$ is the second corner point. The third corner point is the point of intersection between $x = 150$ and $0.8x + 1.0y = 1200$. Substituting $x = 150$ into $0.8x + 1.0y = 1200$, we have the following.

$$0.8(150) + y = 1200 \Rightarrow 120 + y = 1200 \Rightarrow y = 1080$$

Thus, $(150, 1080)$ is the third corner point. The fourth corner point is the point of intersection between $0.8x + 1.0y = 1200$ and $4x + 3y = 4200$. In part (a) this was found to be $(375, 900)$.

We wish to maximize $\$2.65x + \$4.67y$.

Corner Point	Value of the Profit Formula: $\$2.65x + \$4.67y$ (in thousands)						
$(150, 200)$	$\$2.65(150)$	+	$\$4.67(200)$	=	$\$397.50$	+	$\$934.00$	=	$\$1331.50$
$(150, 1080)$	$\$2.65(150)$	+	$\$4.67(1080)$	=	$\$397.50$	+	$\$5043.60$	=	$\$5441.10*$
$(375, 900)$	$\$2.65(375)$	+	$\$4.67(900)$	=	$\$993.75$	+	$\$4203.00$	=	$\$5196.75$
$(900, 200)$	$\$2.65(900)$	+	$\$4.67(200)$	=	$\$2385.00$	+	$\$934.00$	=	$\$3319.00$

Optimal production policy: Make 150 desk lamps and 1080 floor lamps.

For Exercises 45 – 47, part (c) (using a simplex algorithm program) will not be addressed in the solutions.

45. (a) Let x be the number of chairs, y be the number of tables, and z be the number of beds.

	Chis (80 hr)	Sue (200 hr)	Juan (200)	Minimums	Profit
Chairs, x	1	3	2	0	$100
Tables, y	3	5	4	0	$250
Beds, z	5	4	8	0	$350

(b) Profit formula: $P = \$100x + \$250y + \$350z$

Constraints: $x \geq 0$, $y \geq 0$, and $z \geq 0$ (minimums)

$$x + 3y + 5z \leq 80 \,(\text{Chris}); 3x + 5y + 4z \leq 200(\text{Sue}); 2x + 4y + 8z \leq 200\,(\text{Juan})$$

(c) Optimal product policy: Make 50 chairs, 10 tables, and no beds each month for a profit of $7500 in one month.

47. (a) Let w be the number of pounds of Excellent coffee, x be the number of pounds of Southern coffee, y be the number of pounds of World coffee, and z be the number of pounds of Special coffee.

	African (17,600 oz)	Brazilian (21,120 oz)	Columbian (12,320 oz)	Minimums	Profit
Excellent, w pounds	0	0	16	0	$1.80
Southern, x pounds	0	12	4	0	$1.40
World, y pounds	6	8	2	0	$1.20
Special, z pounds	10	6	0	0	$1.00

(b) Profit formula: $P = \$1.80w + \$1.40x + \$1.20y + \$1.00z$

Constraints: $w \geq 0, x \geq 0, y \geq 0$, and $z \geq 0$ (minimums)

$$0w + 0x + 6y + 10z \leq 17,600 \,(\text{African})$$

$$0w + 12x + 8y + 6z \leq 12,120 \,(\text{Brazilian})$$

$$16w + 4x + 2y + 6z \leq 12,320\,(\text{Columbian})$$

(c) Optimal product policy: Make 470 pounds of Excellent, none of Southern, 2400 pounds of World, and 320 pounds of Special for a profit of $4046.

49. Minimize $C = 5x + 11y$ subject to $x \geq 2;\ y \geq 3;\ 3x + y \leq 18;\ 6x + 4y \leq 48$

The corner points are $(2,3),(5,3),(4,6)$, and $(2,9)$.

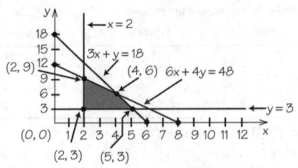

We wish to minimize $C = 5x + 11y$.

Corner Point	Value of the Profit Formula: $5x + 11y$
$(2,3)$	$5(2)\ +\ 11(3)\ =\ 10\ +\ 33\ =\ 43*$
$(5,3)$	$5(5)\ +\ 11(3)\ =\ 25\ +\ 33\ =\ 58$
$(4,6)$	$5(4)\ +\ 11(6)\ =\ 20\ +\ 66\ =\ 86$
$(2,9)$	$5(2)\ +\ 11(9)\ =\ 10\ +\ 99\ =\ 109$

The minimum value occurs at the corner point $(2,3)$, where C is equal to 43.

51. (a) Let x be the number of business calls and y be the charity calls.

	Time (240 min)	Minimums	Profit
Business, x calls	4	0	\$0.50
Charity, y calls	6	0	\$0.40

(b) Profit formula: $P = \$0.50x + \$0.40y$

Constraints: $x \geq 0$ and $y \geq 0$ (minimums); $4x + 6y \leq 240$ (time)

(c) Feasible region:

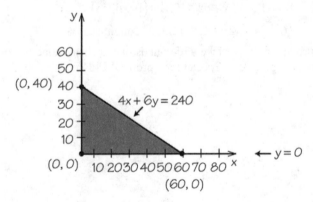

Corner points: The corner points are intercepts on the axes. These are $(0,0)$, $(0,40)$, and $(60,0)$.

Continued on next page

51. continued

(d) We wish to maximize $\$0.50x + \$0.40y$.

Corner Point	Value of the Profit Formula: $\$0.50x + \$0.40y$						
$(0,0)$	$\$0.50(0)$ +	$\$0.40(0)$ =	$\$0.00$ +	$\$0.00$ =	$\$0.00$
$(0,40)$	$\$0.50(0)$ +	$\$0.40(40)$ =	$\$0.00$ +	$\$16.00$ =	$\$16.00$
$(60,0)$	$\$0.50(60)$ +	$\$0.40(0)$ =	$\$30.00$ +	$\$0.00$ =	$\$30.00*$

Optimal production policy: Make 60 business and no charity calls.

With non-zero minimums, the constraints are as follows.

$$x \geq 12 \text{ and } y \geq 10 \text{ (minimums)}; 4x + 6y \leq 240 \text{ (time)}$$

The feasible region looks like the following.

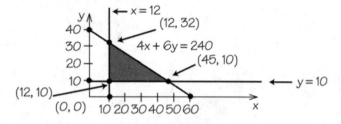

Corner points: One corner point is the point of intersection between $x = 12$ and $y = 10$, namely $(12,10)$. Another is the point of intersection between $x = 12$ and $4x + 6y = 240$. Substituting $x = 12$ into $4x + 6y = 240$, we have the following.

$$4(12) + 6y = 240 \Rightarrow 48 + 6y = 240 \Rightarrow 6y = 192 \Rightarrow y = 32$$

Thus, $(12,32)$ is the second corner point. The third corner point is the point of intersection between $y = 10$ and $4x + 6y = 240$. Substituting $y = 10$ into $4x + 6y = 240$ we have the following.

$$4x + 6(10) = 240 \Rightarrow 4x + 60 = 240 \Rightarrow 4x = 180 \Rightarrow x = 45$$

Thus, $(45,10)$ is the third corner point.

We wish to maximize $\$0.50x + \$0.40y$.

Corner Point	Value of the Profit Formula: $\$0.50x + \$0.40y$						
$(12,10)$	$\$0.50(12)$ +	$\$0.40(10)$ =	$\$6.00$ +	$\$4.00$ =	$\$10.00$
$(12,32)$	$\$0.50(12)$ +	$\$0.40(32)$ =	$\$6.00$ +	$\$12.80$ =	$\$18.80$
$(45,10)$	$\$0.50(45)$ +	$\$0.40(10)$ =	$\$22.50$ +	$\$4.00$ =	$\$26.50*$

Optimal production policy: Make 45 business and 10 charity calls.

53. (a) Let x be the number of bikes and y be the number of wagons.

	Machine (12 hr)	Paint(16 hr.)	Minimums	Profit
Bikes, x	2	4	0	$12
Wagons, y	3	2	0	$10

(b) Profit formula: $P = \$12x + \$10y$

Constraints: $x \geq 0$ and $y \geq 0$ (minimums); $2x + 3y \leq 12$ (machine); $4x + 2y \leq 16$ (paint)

(c) Feasible region:

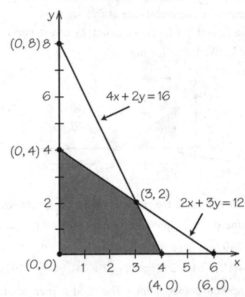

Corner points: Three of the corner points are intercepts on the axes. These are $(0,0)$, $(0,4)$, and $(4,0)$. The final corner point is the point of intersection between $2x + 3y = 12$ and $4x + 2y = 16$. We can find this by multiplying both sides of $2x + 3y = 12$ by -2, and adding the result to $4x + 2y = 16$.

$$-4x - 6y = -24$$
$$\underline{4x + 2y = 16}$$
$$-4y = -8 \Rightarrow y = 2$$

Substitute $y = 2$ into $4x + 2y = 16$ and solve for x. We have the following.

$$4x + 2(2) = 16 \Rightarrow 4x + 4 = 16 \Rightarrow 4x = 12 \Rightarrow x = 3$$

The point of intersection is therefore $(3,2)$.

(d) We wish to maximize $\$12x + \$10y$.

Corner Point	Value of the Profit Formula: $\$12x + \$10y$
$(0,0)$	$\$12(0)$ + $\$10(0)$ = $\$0$ + $\$0$ = $\$0$
$(0,4)$	$\$12(0)$ + $\$10(4)$ = $\$0$ + $\$40$ = $\$40$
$(4,0)$	$\$12(4)$ + $\$10(0)$ = $\$48$ + $\$0$ = $\$48$
$(3,2)$	$\$12(3)$ + $\$10(2)$ = $\$36$ + $\$20$ = $\$56*$

Optimal production policy: Make 3 bikes and 2 wagons.

With non-zero minimums $x \geq 2$ and $y \geq 2$, there will be no change because the optimal production policy already obeys these non-zero minimums.

55. (a)

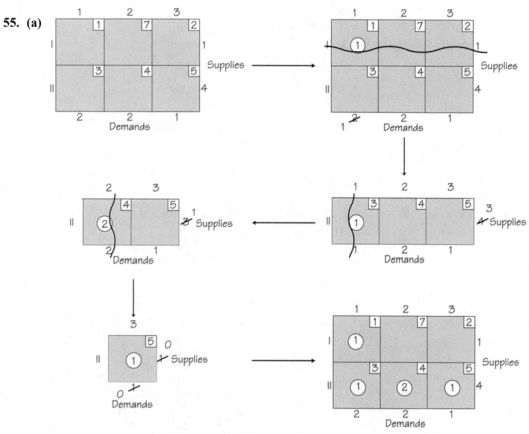

(b) The cost for this solution is $1(1)+1(3)+2(4)+1(5)=1+3+8+5=17.$

(c) The indicator value for cell $(\text{I},2)$ is $7-1+3-4=5.$

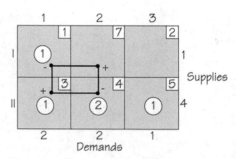

The indicator value for cell $(\text{I},3)$ is $2-1+3-5=-1.$

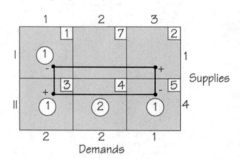

57. (a) The graph is a tree because it is connected and has no circuit.

(b) If we add the edge joining Vertex I to Vertex 2 we get the circuit 2, I, 1, II, 2.

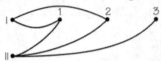

If we add the edge from Vertex I to Vertex 3 we get the circuit 3, I, 1, II, 3.

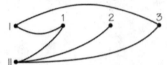

(c) For the circuit 2, I, 1, II, 2 it corresponds to the following circuit of cells.

$$(I,2),(I,1),(II,1),(II,2),(I,2)$$

For the circuit 3, I, 1, II, 3 it corresponds to the following circuit of cells.

$$(I,3),(I,1),(II,1),(II,3),(I,3)$$

59. (a) (i)

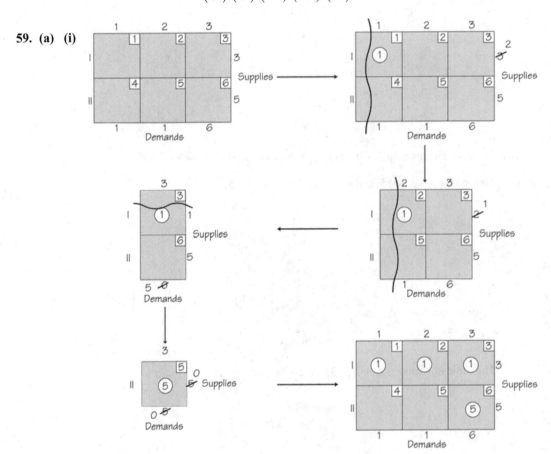

Continued on next page

59. (a) continued

(ii) Since the rim values are the same, the end result will be the same (relative to the new table)

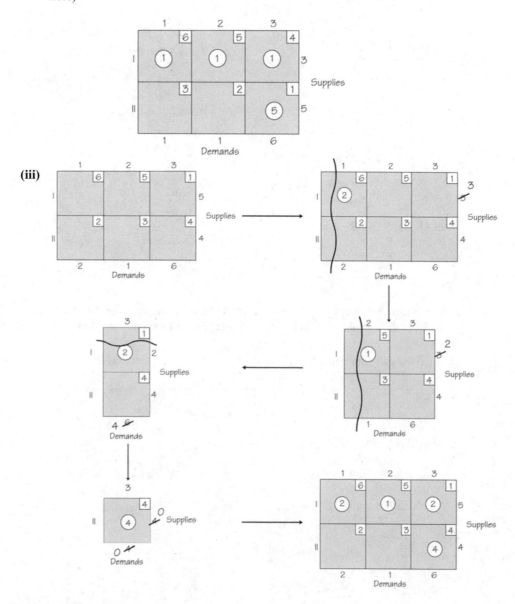

Continued on next page

59. continued

    (b) For (i) the indicator value for each non-circled cell is calculated as follows.

cell $(\text{II},1): 4-6+3-1=0$

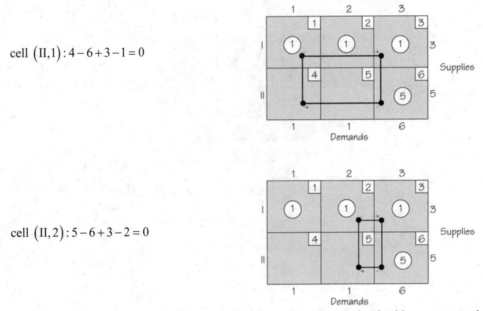

cell $(\text{II},2): 5-6+3-2=0$

The tableau shown is optimal. However, there are also other optimal tableaux, as can be seen from the fact that the indicator values for each of the cells that have no circled entries are 0.

For (ii) the indicator value for each non-circled cell is calculated as follows.

cell $(\text{II},1): 3-1+4-6=0$

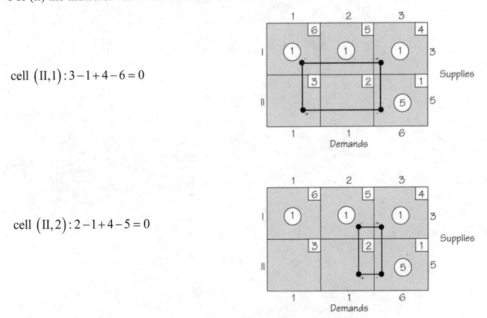

cell $(\text{II},2): 2-1+4-5=0$

The tableau shown is optimal, although there are also other optimal tableaux.

For (iii) the indicator value for each non-circled cell is calculated as follows.

Continued on next page

59. (b) continued

cell $(\text{II},1): 2-4+1-6 = -3$

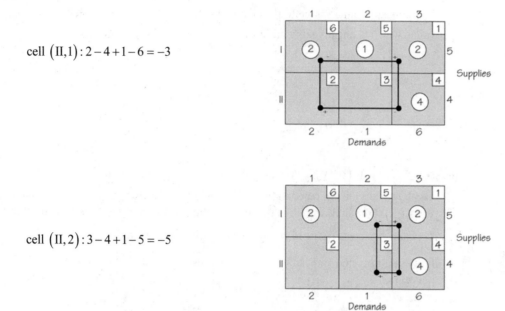

cell $(\text{II},2): 3-4+1-5 = -5$

The tableau shown is not optimal. The current cost is as follows.

$$2(6)+1(5)+2(1)+4(4) = 12+5+2+16 = 35$$

Since cell $(\text{II},1)$ has a more negative indicator value, we can reduce the cost more by using that cell. Increasing by 2, we obtain the following tableau.

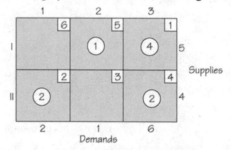

The cost is now $1(5)+4(1)+2(4)+2(2) = 5+4+8+4 = 21$. We can reduce the cost more by using cell $(\text{II},2)$. Increasing by 1, we obtain the following tableau.

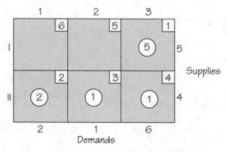

The cost is now $2(2)+1(3)+1(4)+5(1) = 4+3+4+5 = 16$.

61. (i) (a) The cost is given by $4(7)+1(2)+3(1)=28+2+3=33$.

(b) The indicator value for cell $(\mathrm{II},1)$ is $3-1+2-7=-3$. Since this cell has an indicator value, which is negative, the initial solution we found is not optimal.

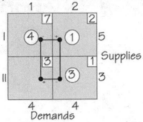

(c) Increasing cell $(\mathrm{II},1)$ by 3 (since this is the minimum of the circled numbers with negative signs in the computation of the indicator) increases the amount shipped via cell $(\mathrm{I},2)$ by 3 and decreases the amount shipped via cell $(\mathrm{II},2)$ and cell $(\mathrm{I},1)$ by 3. This now has cost $4(2)+1(7)+3(3)=8+7+9=24$.

If we multiply the indicator (-3) by 3, this is -9 and $33-24=9$, so the cost is reduced by 9 as expected.

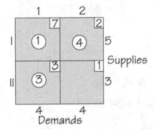

The indicator value for cell $(\mathrm{II},2)$ is
$1-2+7-3=3$.

(ii) (a) The cost is given by $5(6)+1(1)+3(2)=30+1+6=37$.

(b) The indicator value for cell $(\mathrm{I},2)$ is $4-6+1-2=-3$.

Since this cell has an indicator value, which is negative, the initial solution we found is not optimal.

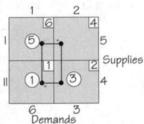

(c) Increasing cell $(\mathrm{I},2)$ by 3 (since this is the minimum of the circled numbers with negative signs in the computation of the indicator) increases the amount shipped via cell $(\mathrm{II},1)$ by 3 and decreases the amount shipped via cell $(\mathrm{II},2)$ and cell $(\mathrm{I},1)$ by 3. This now has cost $3(4)+2(6)+4(1)=12+12+4=28$.

If we multiply the indicator (-3) by 3, this is -9 and $37-28=9$, so the cost is reduced by 9 as expected.

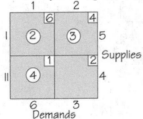

The indicator value for cell $(\mathrm{II},2)$ is
$2-4+6-1=3$.

Chapter 5
Exploring Data: Distributions

Exercise Solutions

1. **(a)** The individuals in the data set are the make and model of the motor vehicles.

    (b) The variables are vehicle type, transmission type, number of cylinders, city MPG, and highway MPG. Histograms would be helpful for cylinders (maybe), and the two MPGs (certainly).

3. Most coins were minted in recent years, so we would expect a peak at the right (highest-numbered years, like 2007 and 2008) and lower bars trailing out to the left of the peak. There are few coins from 1990 and even fewer from 1980, etc. This histogram shape is skewed to the left with a peak at the right and lower bars trailing out to the left of the peak.

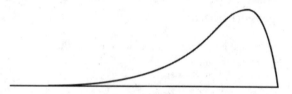

5. **(a)** Otherwise, big countries would top the list even if they had low emissions for their size.

    (b) Using class widths of 2 metric tons per person, we have the following.

Class	Count
0.0 – 1.9	20
2.0 – 3.9	9
4.0 – 5.9	3
6.0 – 7.9	4
8.0 – 9.9	6
10.0 – 11.9	3
12.0 – 13.9	0
14.0 – 15.9	0
16.0 – 17.9	2
18.0 – 19.9	1

The distribution is skewed to the right. There appear to be three high outliers: Canada, Australia, and the United States.

7. **(a)** Alaska is 5.7% and Florida is 17.6%.

    (b) The distribution is single-peaked and roughly symmetric. The center is near 12.7% (12.7% and 12.8% are the 24th and 25th in order out of 48, ignoring Alaska and Florida). The spread is from 8.5% to 15.6%.

9. Here is the stemplot.

10	139
11	5
12	669
13	77
14	08
15	244
16	55
17	8
18	
19	
20	0

There is one high outlier, 200. The center of the 17 observations other than the outlier is 137 (9th of 17). The spread is 101 to 178.

11. **(a)** The repeated stems break up the intervals further. For example the two "3 stems" break the thirties into 30-34 and 35-39.
    (b) The distribution is reasonably symmetric.

13. **(a)** $\bar{x} = \frac{101+103+109+115+126+126+129+137+137+140+148+152+154+154+165+165+178+200}{18} = \frac{2539}{18} \approx 141.1$.

    (b) Without the outlier, $\bar{x} = \frac{101+103+109+115+126+126+129+137+137+140+148+152+154+154+165+165+178}{17} = \frac{2339}{17} \approx 137.6$. The high outlier pulls the mean up.

15. The distribution of incomes is strongly right-skewed, so the mean is much higher than the median. Thus, $66,570 is the mean.

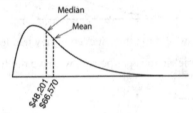

17. Examples will vary.
    One high outlier will do it. For example, 1, 2, 3, 3, 4, 17. This data set has third quartile 4 and mean, $\bar{x} = \frac{1+2+3+3+4+17}{6} = \frac{30}{6} = 5$. Another example is 1, 2, 2, 2, 3, 3, 4, 17. This data set has third quartile $\frac{3+4}{2} = \frac{7}{2} = 3.5$ and mean $\bar{x} = \frac{1+2+2+2+3+3+4+17}{8} = \frac{34}{8} = 4.25$.

19. The five-number summary is Minimum, Q_1, M, Q_3, Maximum.
    The minimum is 5.7 and maximum is 17.6. The median is the mean of the 25th and 26th pieces of data, namely $\frac{12.7+12.8}{2} = \frac{25.5}{2} = 12.75$. There are 25 pieces of data below the median, thus Q_1 is the 13th piece of data, 11.7. There are 25 pieces of data above the median, thus Q_3 is the 38th piece of data, namely 13.5. Thus, the five-number summary for these 50 observations is 5.7, 11.7, 12.75, 13.5, 17.6.

21. The minimum is 5799 and the maximum is 36,700. Since there are 55 pieces of data, the median is the mean of the 28th piece of data, namely 25,942. Since there are 27 observations to the left of the median, Q_1 is 14th piece of data, namely 20,000. Since there are 27 observations to the right of the median, Q_3 is the 42nd piece of data, namely 34,986.

Thus, the five-number summary is 5799, 20000, 25942, 34986, 36700. The boxplot displays these five numbers.

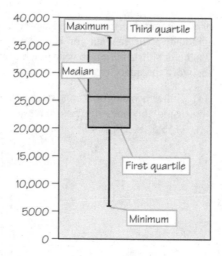

Distinctive features of the histogram missed could be that according to the histogram there are no tuition and fees between $10,000 and $14,000 (thereby creating two distinct clusters of values). Also, the most frequent tuition and fees is $36,000.

23. To determine the minimum, maximum, and median, we should put the 48 pieces of data in order from smallest to largest. It may be easier, however, to create a stemplot.

```
 0 | 001122223357899
 1 | 02478
 2 | 3558
 3 | 67899
 4 | 68
 5 | 1
 6 | 18
 7 | 36
 8 | 018
 9 | 017
10 | 02
11 | 0
12 |
13 |
14 |
15 |
16 | 0
17 | 0
18 |
19 | 9
```

The minimum is 0.0 and the maximum is 19.9. Since there are 48 pieces of data, the median is the mean of the 24th and 25th pieces of data, namely $\frac{2.8+3.6}{2} = \frac{6.4}{2} = 3.2$. Since there are 24 observations to the left of the median, Q_1 is the mean of the 12th and 13th piece, of data, namely $\frac{0.7+0.8}{2} = \frac{1.5}{2} = 0.75$. Since there are 24 observations to the right of the median, Q_3 is the mean of the 36th and 37th pieces of data, namely $\frac{7.6+8.0}{2} = \frac{15.6}{2} = 7.8$.

Thus, the five-number summary is 0.0, 0.75, 3.2, 7.8, 19.9. The third quartile and maximum are much farther from the median than the first quartile and minimum, showing that the right side of the distribution is more spread out than the left side.

25. The income distribution for bachelor's degree holders is generally higher than for high school graduates: the median for bachelor's is greater than the third quartile for high school. The bachelor's distribution is very much more spread out, especially at the high-income end but also between the quartiles.

27. **(a)** Either a histogram or a stemplot will do. By rounding to the nearest whole number and using the ones digit as leaves, we have the following stemplot, with three high outliers (156.5, 196.0, 204.9) omitted.

```
 0 | 2337
 1 | 0022355889
 2 | 0125789
 3 | 1133455577889
 4 | 3355689
 5 | 0234679
 6 | 1356
 7 | 00
 8 | 012
 9 | 28
10 | 3
11 | 8
```

Here is the histogram using class widths of 20 (thousand) barrels of oil.

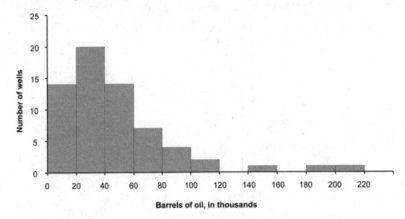

The distribution is right-skewed with high outliers.

(b) $\bar{x} = \frac{3087.9}{64} \approx 48.25$ and since there are 64 pieces of data, the median is the mean of the 32nd and 33rd pieces of data, namely $M = \frac{37.7+37.9}{2} = \frac{75.6}{2} = 37.8$. The long right tail pulls the mean up.

(c) The minimum is 2.0 and the maximum is 204.9. As found in part b, the median is 37.8. Since there are 32 observations to the left of the median, Q_1 is the mean of the 16th and 17th piece of data, namely $\frac{21.3+21.7}{2} = \frac{43}{2} = 21.5$. Since there are 32 observations to the right of the median, Q_3 is the mean of the 48th and 49th piece of data, namely $\frac{58.8+61.4}{2} = \frac{120.2}{2} = 60.1$.

Thus, the five-number summary is 2.0, 21.5, 37.8, 60.1, 204.9. The third quartile and maximum are much farther above the median that the first quartile and minimum are below it, showing that the right side of the distribution is much more spread out than the left side.

29. For the data in Table 5.1, $M = 4.7\%$, $Q_1 = 2.1\%$, and $Q_3 = 8.7\%$. So $IQR = 8.7 - 2.1 = 6.6$ and $1.5IQR = 1.5(6.6) = 9.9$. Values less than $2.1 - 9.9 = -7.8$ or greater than $8.7 + 9.9 = 18.6$ are suspected outliers. By this criterion, there are 5 high outliers: Arizona, California, Nevada, New Mexico, and Texas.

31. (a) Placing the data in order (not required, but helpful), we have the following hand calculations.

Observations x_i	Deviations $x_i - \bar{x}$	Squared deviations $(x_i - \bar{x})^2$
4.88	− 0.57	0.3226
5.07	− 0.38	0.1429
5.10	− 0.35	0.1211
5.26	− 0.19	0.0353
5.27	− 0.18	0.0317
5.29	− 0.16	0.0250
5.29	− 0.16	0.0250
5.30	− 0.15	0.0219
5.34	− 0.11	0.0117
5.34	− 0.11	0.0117
5.36	− 0.09	0.0077
5.39	− 0.06	0.0034
5.42	− 0.03	0.0008
5.44	− 0.01	0.0001
5.46	0.01	0.0001
5.47	0.02	0.0005
5.50	0.05	0.0027
5.53	0.08	0.0067
5.55	0.10	0.0104
5.57	0.12	0.0149
5.58	0.13	0.0174
5.61	0.16	0.0262
5.62	0.17	0.0296
5.63	0.18	0.0331
5.65	0.20	0.0408
5.68	0.23	0.0538
5.75	0.30	0.0912
5.79	0.34	0.1170
5.85	0.40	0.1616
sum = 157.99	**sum = 0.00**	**sum = 1.3669**

$\bar{x} = \frac{157.99}{29} \approx 5.448$ and $s^2 = \frac{1.3669}{29-1} = \frac{1.3669}{28} \approx 0.0488$ which implies $s \approx \sqrt{0.0488} \approx 0.221$.

(b) The median is the $\frac{29+1}{2} = \frac{30}{2} = 15^{\text{th}}$ piece of data, namely 5.46 (From Exercise 10). The mean and median are close, but the one low observation pulls $\bar{x}$ slightly below M.

33. For Data A:

Placing the data in order (not required, but helpful), we have the following hand calculations.

Observations	Deviations	Squared deviations
x_i	$x_i - \bar{x}$	$(x_i - \bar{x})^2$
3.10	-4.40	19.3600
4.74	-2.76	7.6176
6.13	-1.37	1.8769
7.26	-0.24	0.0576
8.10	0.60	0.3600
8.14	0.64	0.4096
8.74	1.24	1.5376
8.77	1.27	1.6129
9.13	1.63	2.6569
9.14	1.64	2.6896
9.26	1.76	3.0976
sum = 82.51	**sum = 0.01**	**sum = 41.2763**

$\bar{x} = \frac{82.51}{11} \approx 7.50$ and $s^2 = \frac{41.2763}{11-1} = \frac{41.2763}{10} = 4.12763$ and $s = \sqrt{4.12763} \approx 2.03$.

Note: The sum of the deviations is 0.01 due to rounding off the mean prior to finding the deviations.

Using the TI-83, we have the following.

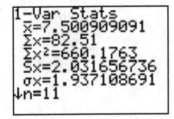

$\bar{x} \approx 7.50$ and $s \approx 2.03$.

Continued on next page

33. continued
For Data B:

Placing the data in order (not required, but helpful), we have the following hand calculations.

Observations x_i	Deviations $x_i - \bar{x}$	Squared deviations $\left(x_i - \bar{x}\right)^2$
5.39	-2.11	4.4521
5.73	-1.77	3.1329
6.08	-1.42	2.0164
6.42	1.08	1.1664
6.77	-0.73	0.5329
7.11	-0.39	0.1521
7.46	-0.04	0.0016
7.81	0.31	0.0961
8.15	0.65	0.4225
8.84	1.34	1.7956
12.74	5.24	27.4576
sum = 82.5	**sum = 0.00**	**sum = 41.2262**

$\bar{x} = \frac{82.50}{11} = 7.50$ and $s^2 = \frac{41.2262}{11-1} = \frac{41.2262}{10} = 4.12262$ and $s = \sqrt{4.12262} \approx 2.03$.

Using the TI-83, we have the following.

```
1-Var Stats
x̄=7.5
Σx=82.5
Σx²=659.9762
Sx=2.030423601
σx=1.935932944
↓n=11
```

$\bar{x} = 7.50$ and $s \approx 2.03$.

For both data, $\bar{x} = 7.50$ and $s = 2.03$ (to two decimal places).

Rounding Data A to the nearest tenth, we have the following.

9.1   8.1   8.7   8.8

9.3   8.1   6.1   3.1

9.1   7.3   4.7

Data A has two low outliers:

3	1
4	7
5	
6	1
7	3
8	1178
9	113

Rounding Data B to the nearest tenth, we have the following.

7.5   6.8   12.7   7.1

7.8   8.8   6.1   5.4

8.2   6.4   5.7

Data B has one high outlier:

5	5
6	148
7	1578
8	28
9	
10	
11	
12	7

Data A have two low outliers and Data B have one high outlier. Additional comments may vary.

35. (a) $s = 0$ is smallest possible: 1, 1, 1, 1

  (b) Largest possible spread: 0, 0, 10, 10.

  (c) In part (a), the answer is not unique. Any other set of four identical numbers will yield a standard deviation of 0, i.e. the values do not deviate from the mean, which is that repeated number.

  (d) In part (b), the answer is unique. The data are spread out as much as possible, given the constraints.

37. Approximately 68% of the students will receive a grade of C. Approximately $\left(\dfrac{95-68}{2}\right)\% = \dfrac{27}{2}\% = 13.5\%$ will receive a grade of B. Thus, $0.135(200) = 27$ will receive a grade of B.

39. Left-skewed, so the mean is pulled toward the long left tail: A = mean and B = median.

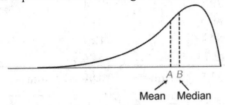

41. (a) $\mu \pm 3\sigma = 336 \pm 3(3) = 336 \pm 9$, or 327 to 345 days.

  (b) Make a sketch: 339 days is one σ above μ; 68% are within σ of μ.

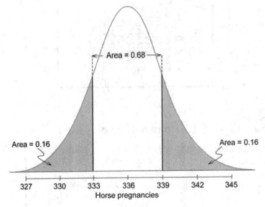

32% lie farther from μ. Thus, half of these, or 16%, lie above 339.

43. The quartiles are $\mu \pm 0.67\sigma = 1511 \pm 0.67(194) \approx 1511 \pm 130$, or $Q_1 = 1381$ and $Q_3 = 1641$.

45.

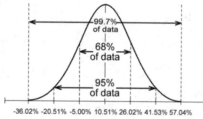

-36.02% -20.51% -5.00% 10.51% 26.02% 41.53% 57.04%

Yearly stock return

(a) $\mu \pm 2\sigma = 10.51 \pm 2(15.51) = 10.51 \pm 31.02$, or -20.51% to 41.53%.

(b) A loss of 20.51% or greater.

47. (a) Normal curves are symmetric, so median = mean = 10%.

(b) Because 95% of values lie within 2σ of μ, $\mu \pm 2\sigma = 10 \pm 2(0.2) = 10 \pm 0.4$ implies 9.6% to 10.4% is the range of concentrations that cover the middle 95% of all the capsules.

(c) The range between the two quartiles covers the middle half of all capsules. Thus, $\mu \pm 0.67\sigma = 10 \pm 0.67(0.2) = 10 \pm 0.134$ implies 9.866% to 10.134% is the desired range.

49.

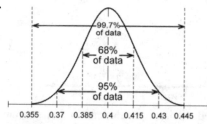

0.355   0.37   0.385   0.4   0.415   0.43   0.445

Portion of people who stay home
for fear of crime

(a) 50% above 0.4, because of the symmetry of normal curves; 0.43 is 2σ above μ, so 2.5%.

(b) $\mu \pm 2\sigma = 0.4 \pm 2(0.015) = 0.4 \pm 0.03$, or 0.37 to 0.43.

51. Note: Values are rounded to the nearest tenth.
Lengths of red flowers are somewhat right-skewed, with no outliers:

```
37 | 489
38 | 00112289
39 | 268
40 | 67
41 | 5799
42 | 02
43 | 1
```

Lengths of yellow flowers are quite symmetric, with no outliers:

```
34 | 66
35 | 257
36 | 0015788
37 | 01
38 | 1
```

53. Red:

Observations x_i	Deviations $x_i - \overline{x}$	Squared deviations $\left(x_i - \overline{x}\right)^2$
37.40	− 2.311304	5.34213
37.78	− 1.931304	3.72994
37.87	− 1.841304	3.39040
37.97	− 1.741304	3.03214
38.01	− 1.701304	2.89444
38.07	− 1.641304	2.69388
38.10	− 1.611304	2.59630
38.20	− 1.511304	2.28404
38.23	− 1.481304	2.19426
38.79	− 0.921304	0.84880
38.87	− 0.841304	0.70779
39.16	− 0.551304	0.30394
39.63	− 0.081304	0.00661
39.78	0.068696	0.00472
40.57	0.858696	0.73736
40.66	0.948696	0.90002
41.47	1.758696	3.09301
41.69	1.978696	3.91524
41.90	2.188696	4.79039
41.93	2.218696	4.92261
42.01	2.298696	5.28400
42.18	2.468696	6.09446
43.09	3.378696	11.41559
sum = **913.36**	sum = **0.000008**	sum = **71.18206**

$\overline{x} = \frac{913.36}{23} \approx 39.71$ (we used $\overline{x} \approx 39.711304$ in the deviations calculations for better accuracy and rounded to five decimal places in the calculation of squared deviations). We therefore have $s^2 \approx \frac{71.18206}{23-1} = \frac{71.18206}{22} \approx 3.2355$ which implies $s \approx \sqrt{3.2355} \approx 1.799$.

Yellow:

Observations x_i	Deviations $x_i - \overline{x}$	Squared deviations $\left(x_i - \overline{x}\right)^2$
34.57	− 1.61	2.5921
34.63	− 1.55	2.4025
35.17	− 1.01	1.0201
35.45	− 0.73	0.5329
35.68	− 0.50	0.2500
36.03	− 0.15	0.0225
36.03	− 0.15	0.0225
36.11	− 0.07	0.0049
36.52	0.34	0.1156
36.66	0.48	0.2304
36.78	0.60	0.3600
36.82	0.64	0.4096
37.02	0.84	0.7056
37.10	0.92	0.8464
38.13	1.95	3.8025
sum = **542.70**	sum = **0.00**	sum = **13.3176**

$\overline{x} = \frac{542.70}{15} = 36.18$ and $s^2 = \frac{13.31766}{15-1} = \frac{13.31766}{14} \approx 0.9513$ which implies $s \approx \sqrt{0.9513} \approx 0.975$.

The mean and standard deviation are better suited to the symmetrical yellow distribution.

55. The top 2.5% of the distribution lies above
$$36.18 + 2(0.975) = 38.13 \text{ millimeters.}$$
The top 16% of the distribution lies above
$$36.18 + 1(0.975) = 37.155 \text{ millimeters.}$$
The top 25% of the distribution lies above
$$36.18 + 0.67(0.975) = 36.83 \text{ millimeters.}$$

The value 37.4 is between 37.155 and 38.13, so between 2.5% and 16% of yellow flowers are longer that 37.4 millimeters.

57. If every number in a data set is increased by 10, then the mode, mean, median will each increase by 10. The range and standard deviation, however, will not change.

Chapter 6
Exploring Data: Relationships

Exercise Solutions

1. **(a)** It is more reasonable to explore study time as an explanatory variable and the exam grade as the response variable.

   (b) It is more reasonable to explore the relationship only.

   (c) It is more reasonable to explore rainfall as an explanatory variable and the corn yield as the response variable.

   (d) It is more reasonable to explore the relationship only.

3. **(a)** Life expectancy increases with GDP in a curved pattern. The increase is very rapid at first, but levels off for GDP above roughly $5000 per person.

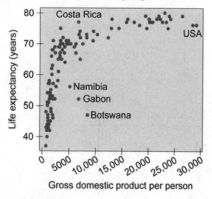

   (b) Richer nations have better diets, clean water, and better health care, so we expect life expectancy to increase with wealth. But once food, clean water, and basic medical care are in place, greater wealth has only a small effect on lifespan.

5. The scatterplot is as follows.

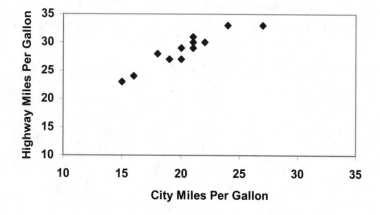

Continued on next page

5.   continued
Using a TI-83, we get the following.

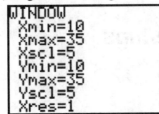 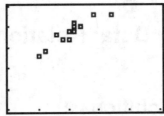

Purists should notice that because the variables measure similar quantities the plot is square with the same scales on both axes. There is a strong positive straight-line relationship.

7.   **(a)** The speed is the explanatory variable.

(b) The scatterplot is as follows.

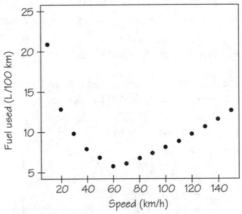

Using a TI-83, we get the following.

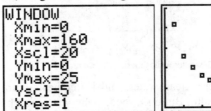

The relationship is curved; fuel usage first decreases as speed increases (higher gears cover more ·distance per motor revolution) then increases as speed is further increased (air resistance builds at higher speeds).

(c) There is no overall direction.

(d) The relationship is quite strong. There is little scatter about the overall curved pattern.

9.   The estimated slope would be $\dfrac{506-386}{191-139} = \dfrac{120}{52} \approx 2.31$.

11. (a) Choose two values of weeks, preferably near 1 and 150. Find pH from the equation given, plot the two points (weeks horizontal) and draw the line through them.

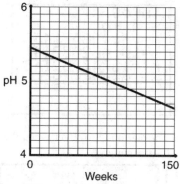

Using a TI-83, we get the following.

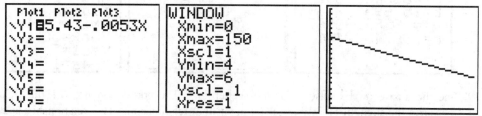

(b) Week 1: predicted pH $= 5.43 - (0.0053 \times 1) = 5.43 - 0.0053 = 5.4247 \approx 5.42$

Week 150: predicted pH $= 5.43 - (0.0053 \times 150) = 5.43 - 0.795 = 4.635 \approx 4.64$

(c) The slope -0.0053 says that on average pH declined by 0.0053 per week during the study period.

13. Some sample ages would be as follows.

Age of Woman	Age of Man
18	20
20	22
27	29
32	34
41	43

Using a TI-83, we get the following.

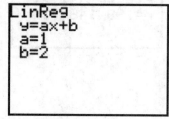

Using the linear regression feature, we obtain the following.

```
LinReg
y=ax+b
a=1
b=2
```

The line is $y = x + 2$, and the slope would be 1.

15. With a correlation of 0.9353, the indication is a very strong straight-line pattern. This is consistent with the scatterplot of the data.

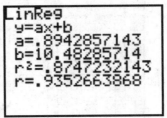

17. The correlation is 0.9674. The correlation is stronger when the Prius is added, because that point extends (strengthens) the straight-line pattern.

19. See the answer to Exercise 7 for the scatterplot. The correlation is −0.1700. Correlation measures the strength of only linear (straight-line) relationship. This relationship is strong but curved (not linear).

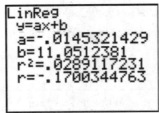

21. The correlation would be 1 because there is a perfect straight-line relationship, $y = x + 3$.

23. **(a)** Negative: Older cars will in general sell for lower prices.

  (b) Negative: Heavier cars will (other things being equal) get fewer miles per gallon.

  (c) Positive: Taller people are on average heavier than shorter people.

  (d) Small: There is no reason to expect that height and IQ are related.

25. Ask how similar the market sector of each fund is to large U.S. stocks and arrange in order.

  (a) Dividend Growth, $r = 0.98$; Small Cap Stock, $r = 0.81$; Emerging Markets, $r = 0.35$.

  (b) No: it just says that they tend to move in the same direction, whether up or down.

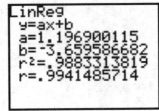

27. **(a)** Predicted highway mpg $= 10.48 + 0.89(\text{city mpg})$.

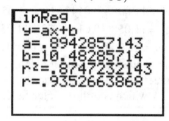

(b) Predicted highway mpg $= 10.48 + 0.89(17) = 10.48 + 15.13 = 25.61$. Thus, you would predict 25.61 mpg.

(c) The points in the plot follow a straight line, so we expect quite accurate predictions, especially within the range of the x values already observed.

29. Choose two city mpg values, such as $x = 10$ and $x = 30$, and use the equation of the line to find each value of y. Plot the two points and draw the line between them. Here is the plot.

$x = 10$: Predicted Highway mpg $= 10.48 + 0.89(10) = 10.48 + 8.9 = 19.38$ MPG.

$x = 30$: Predicted Highway mpg $= 10.48 + 0.89(30) = 10.48 + 26.7 = 37.18$ MPG.

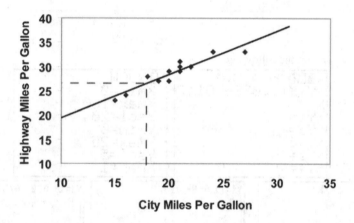

Predicted highway mileage of a car that gets 18 mpg in the city is approximately 26 mpg or 27 mpg (26.5 mph).

Using a TI-83, we get the following.

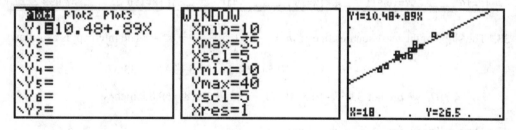

31. Since predicted fuel $= 11.058 - 0.0147 \times$ speed, we have the following.

Speed = 10 km/h: predicted fuel $= 11.058 - 0.0147 \times 10 = 11.058 - 0.147 = 10.911$ L/100km

Speed = 70 km/h: predicted fuel $= 11.058 - 0.0147 \times 70 = 11.058 - 1.029 = 10.029$ L/100km

Speed = 150 km/h: predicted fuel $= 11.058 - 0.0147 \times 150 = 11.058 - 2.205 = 8.853$ L/100km

The predicted values from the equation given are approximately 10.91, 10.03, and 8.85, respectively. The observed values are 21.00, 6.30, and 12.88, respectively. The least-squares line gives the best straight-line fit, which is of little value here.

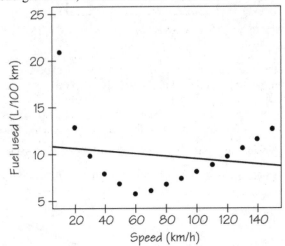

Using a TI-83, we get the following.

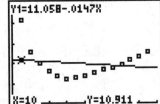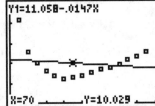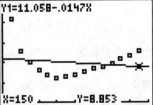

33. The slope of the least-squares line is $m = r \dfrac{s_y}{s_x} = 0.5 \left(\dfrac{2.7}{2.5} \right) = 0.54$. The intercept is as follows.

$$b = \bar{y} - m\bar{x} = 68.5 - (0.54)(64.5) = 68.5 - 34.83 = 33.67$$

For $x = 67$, we predict $33.67 + (0.54)(67) = 33.67 + 36.18 = 69.85$ inches.

35. The predicted y for $x = \bar{x}$ is as follows.

$$\hat{y} = m\bar{x} + b = \left(r \dfrac{s_y}{s_x} \right) \bar{x} + \left(\bar{y} - m\bar{x} \right) = \left(r \dfrac{s_y}{s_x} \right) \bar{x} + \bar{y} - \left(r \dfrac{s_y}{s_x} \right) \bar{x} = \bar{y}$$

37. First compare the distributions for the two years. To make the boxplots, we need the five-number summary for each data set.

Putting the 2002 data in order, we have the following.

$$-50.5, -49.5, -47.8, -42, -37.8, -26.9, -23.4, -21.1, -18.9, -17.2, -17.1,$$
$$-12.8, -11.7, -11.5, -11.4, -9.6, -7.7, -6.7, -5.6, -2.3, -0.7, -0.7, 64.3$$

The minimum is -50.5 and the maximum is 64.3. The median is the $\frac{23+1}{2} = \frac{24}{2} = 12^{\text{th}}$ piece of data, namely -12.8. Since there are 11 observations to the left of the median, Q_1 is the $\frac{11+1}{2} = \frac{12}{2} = 6^{\text{th}}$ piece of data, namely -26.9. Since there are 11 observations to the right of the median, Q_3 is the $12 + 6 = 18^{\text{th}}$ piece of data, namely -6.7.

Thus, the five - number summary is $-50.5, -26.9, -12.8, -6.7, 64.3$.

Putting the 2003 data in order, we have the following.

$$14.1, 19.1, 22.9, 23.9, 26.1, 27.5, 28.7, 29.5, 30.6, 31.1, 32.1,$$
$$32.3, 35.0, 36.5, 36.9, 36.9, 41.8, 43.9, 57.0, 59.4, 62.7, 68.1, 71.9$$

The minimum is 14.1 and the maximum is 71.9. The median is the $\frac{23+1}{2} = \frac{24}{2} = 12^{\text{th}}$ piece of data, namely 32.3. Since there are 11 observations to the left of the median, Q_1 is the $\frac{11+1}{2} = \frac{12}{2} = 6^{\text{th}}$ piece of data, namely 27.5. Since there are 11 observations to the right of the median, Q_3 is the $12 + 6 = 18^{\text{th}}$ piece of data, namely 43.9.

Thus, the five - number summary is $14.1, 27.5, 32.3, 43.9, 71.9$.

Here are boxplots.

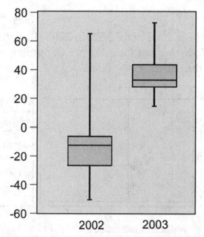

Using a TI-83, we get the following. The top boxplot is for 2002 and the bottom one is for 2003.

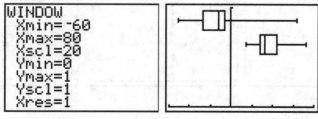

Continued on next page

37. continued

Histograms (or stemplots) show that the 2003 returns are roughly single-peaked and symmetric, and that the 2002 returns are left-skewed with an extreme high outlier. Below are the histograms.

Class	Count
− 60 − (− 40.1)	4
− 40 − (− 20.1)	4
− 20 − (− 0.1)	14
0 − 19.9	0
20 − 39.9	0
40 − 59.9	0
60 − 79.9	1

Class	Count
0 − 19.9	2
20 − 39.9	14
40 − 59.9	4
60 − 79.9	3

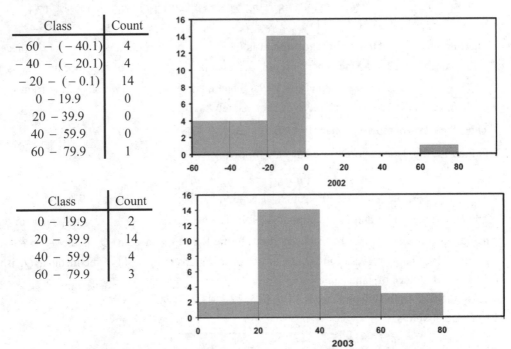

The median returns are − 12.8% in 2002 and 32.3% in 2003 (from the five - number summary). The correlation is $r = − 0.616$; because of the influence of outliers on correlation, it is better to report the correlation without the outlier, $r = − 0.838$.

With Outlier

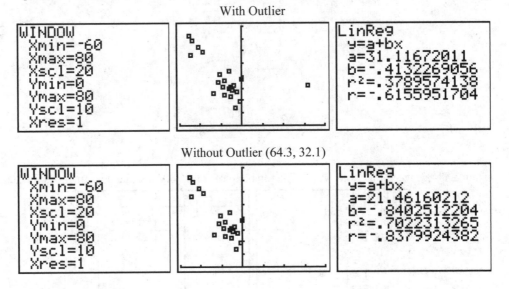

Without Outlier (64.3, 32.1)

Continued on next page

37. continued

That is, the funds that went down most in 2002 tended to go up most in 2003. The scatterplot confirms this:

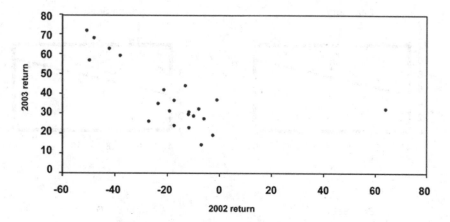

39. (a) All four sets of data have $r = 0.816$ and regression line $y = 3.0 + 0.5x$ to a close approximation. (Note: Screenshots are shown for $y = a + bx$)

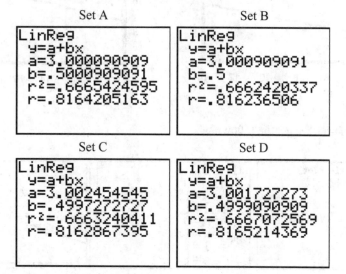

Continued on next page

39. continued

  (b) Here are the plots:

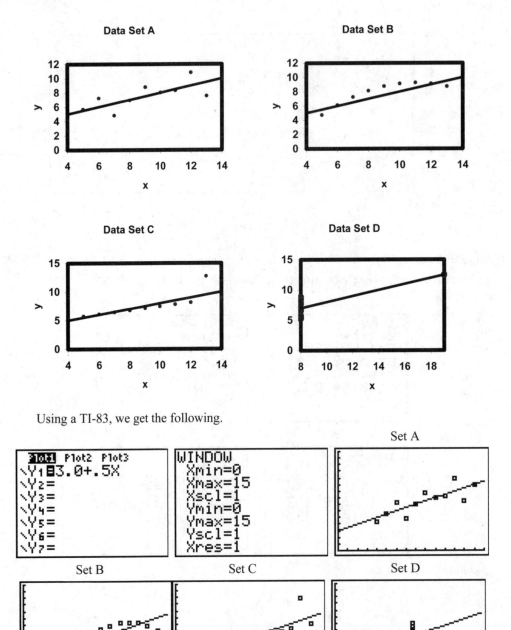

Using a TI-83, we get the following.

  (c) Only A is a "normal" regression setting in which the line is useful for prediction. Plot B is curved, C has an extreme outlier in y, and D has all but one x identical. The lesson: Plot your data before calculating.

In Exercises 41 & 43, answers will vary.

41. Heavier people who are concerned about their weight may be more likely than lighter people to choose artificial sweeteners in place of sugar.

43. Higher income generally means better water and sewage utilities, better diet, and better medical care, which will produce better health. But better health means more children can go to school and more workers are able to work and can stay on the job, which raises national income. For example, AIDS is having a direct negative effect on the economies of African nations.

45. Lead level is the explanatory variable and reading score is the response variable. Because of the dangers of lead exposure, we expect any association to be negative, and the scatterplot confirms this negative association.

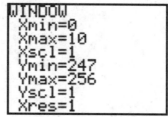 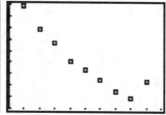

47. (a) There is a positive association, so r will be positive. The pattern is a bit irregular, so r won't be close to 1.

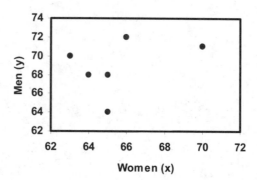

Using a TI-83, we get the following.

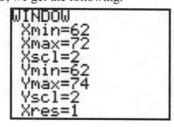 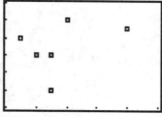

(b) $r = 0.3602$.

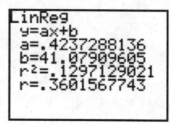

49. (a) The slope is $m = 0.42$. For each additional inch of women's height, the height of the next person dated goes up by 0.42 inch on average.

(b) The prediction is as follows.

$$\text{predicted height of date} = 41.08 + 0.42 \times \text{female height}$$
$$= 41.08 + 0.42(67) = 41.08 + 28.14 = 69.22 \text{ inches}$$

Chapter 7
Data for Decisions

Exercise Solutions

1. **(a)** Population: U.S. residents aged 18 and older.

   (b) Sample: The 1,027 who responded.

3. Students passing the student center may not fairly represent all students. For example, they may underrepresent commuters or students whose classes are far from the center. In addition, a woman student may be reluctant to stop men, thus underrepresenting male students.

5. **(a)** This isn't clear: possibly its readers, possibly all adults in its circulation area.

   (b) Larger; People with strong opinions, especially negative opinions, are more likely to respond. This is bias due to voluntary response.

7. If we assign labels 01 to 20 to the girls' names and start at line 117 in Table 7.1, our sample is 16 = Taylor, 18 = Brianna, and 06 = Alexis.

9. **(a)** 001 to 371.

   (b) Area codes labeled 214, 235, 119.

11. If you always start at the same point in the table, your sample is predictable in advance. Repeated samples of the same size from the same population will always be the same - that's not random.

13. **(a)** Because $\frac{200}{5} = 40$ we divide the list into 5 groups of 40. (By the way, if the list has 204 rooms, we divide it into 5 groups of 40 and final group of 4. A sample contains a room from the final group only when the first room chosen is among the first 4 in the list.) Label the first 40 rooms 01 to 40. Line 120 chooses room 35. The sample is rooms 35, 75, 115, 155, and 195.

    (b) Each of the first 40 rooms has chance 1 in 40 to be chosen. Each later room is chosen exactly when the corresponding room in the first 40 is chosen. So every room has equal chance, 1 in 40. The only possible samples consist of 5 rooms spaced 40 apart in the list. An SRS gives *all* samples of 5 rooms an equal chance to be chosen.

15. **(a)** All people aged 18 and over living in the United States.

    (b) Of the 1334 called, 403 did not respond. The rate is $\frac{403}{1334} \approx 0.30 = 30\%$.

    (c) It is hard to remember exactly how many movies you saw in exactly the past 12 months.

17. Answers will vary.
    People are more reluctant to "change" the Constitution than to "add to" it. So the wording "adding to" will produce a higher percent in favor.

19. No treatment was imposed on the subjects. This observational study collected unusually detailed information about the subjects but made no attempt to influence them.

21. Answers will vary.

    (a) It is an observational study that gathers information (e.g., through interviews) without imposing any treatment.

    (b) "Significant" means "unlikely to be due simply to chance."

    (c) Nondrinkers might be more elderly or in poorer health than moderate drinkers.

23. The design resembles Figure 7.3. Be sure to show randomization, two groups and their treatments, and the response variable (change in obesity).

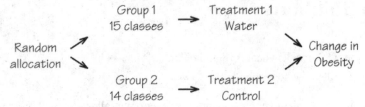

If we label the 29 classes 01 to 29 and choose 15 for the treatment group, this group contains classes 17, 09, 22, 13, 07, 02, 27, 01, 18, 25, 29, 19, 14, 15, 08. We used lines 103 to 106 of Table 7.1, skipping many duplicate pairs of digits. The remaining 14 classes make up the control group.

25. This is a randomized comparative experiment with four branches, similar to Figure 7.4 with one more branch. The "flow chart" outline must show random assignment of subjects to groups, the four treatments, and the response variable (health care spending).

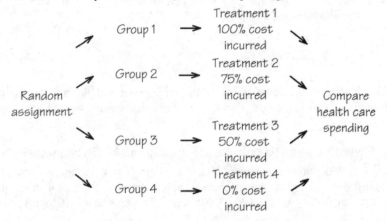

We can't show the group sizes because we don't know how many people or households are available to participate.

27. **(a)** There are 6 treatments, each combination of a level of discount and fraction on sale. In table form, the treatments are

	Discount level		
	20%	40%	60%
50% on sale	1	2	3
100% on sale	4	5	6

(b) The outline randomly assigns 10 students to each of the 6 treatment groups, then compares the attractiveness ratings. It resembles Figure 7.3, but with 6 branches.

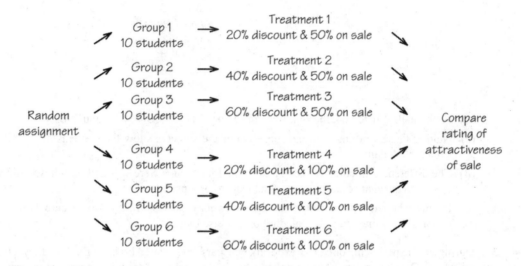

Label the subjects 01 to 60 and read line 123 of Table 7.1. The first group contains subjects labeled 54, 58, 08, 15, 07, 27, 10, 25, 60, 55.

29. **(a)** The design resembles Figure 7.3:

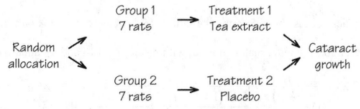

(b) Label the rats 01 to 14. The tea group contains 07, 09, 06, 08, 12, 04, 11.

31. (a) This is a randomized comparative experiment with four branches. The "flow chart" outline must show random assignment of subjects to groups, the group sizes and treatments, and the response variable (colon cancer). It is best to use groups of equal size, 216 people in each group.

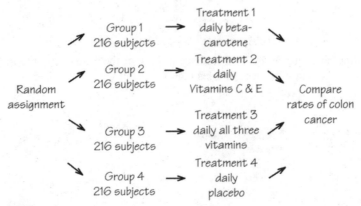

(b) With labels 001 to 864, the first five chosen are 731, 253, 304, 470, and 296.

(c) Neither those working with the subjects nor the subjects know the contents of the pill each subject took daily.

(d) The differences in colon cancer cases in the four groups were so small that they could easily be due to the chance assignment of subjects to groups.

(e) People who eat lots of fruits and vegetables may eat less meat or more cereals than other people. They may drink less alcohol or exercise more.

33. During the experiment, only the experimental cars had center brake lights, so they attracted attention. Once most cars had them, they were less noticed and so did a poorer job of preventing collisions. This is an example of an experiment that could not be completely realistic.

35. This is an observational study. There was a measurement of information, but there was no attempt to influence the response.

37. Both are statistics because both describe the sample (the subjects who took part in the study).

39. (a) The distribution is approximately normal with mean $p = 0.14$ and a standard deviation as follows.

$$\sqrt{\frac{p(1-p)}{n}} = \sqrt{\frac{0.14(1-0.14)}{500}} = \sqrt{\frac{0.14(0.86)}{500}} \approx 0.0155$$

(b) $0.14 \pm (2)(0.0155) = 0.14 \pm 0.031$

$$0.14 - 0.031 = \underline{0.109} \text{ to } 0.14 + 0.031 = \underline{0.171}$$

41. (a) Each digit in the table has one chance in 10 to be any of the ten possible digits 0, 1, 2, 3, 4, 5, 6, 7, 8, 9. So in the long run, 60% of the digits we encounter will be 0, 1, 2, 3, 4, or 5 and 40% will be 6, 7, 8, or 9.

(b) Line 101 contains 29 digits 0 to 5. This stands for a sample with $\frac{29}{40} = 0.725 = 72.5\%$ "yes" responses. If we use lines 101 to 110 to simulate ten samples, the counts of "yes" responses are 29, 24, 23, 23, 20, 24, 23, 19, 24, and 18. Thus, three samples are exactly correct $\left(\frac{24}{40} = 0.60 = 60\%\right)$, one overestimates, and six underestimate.

43. The sample proportion who claim to have attended is $\hat{p} = \frac{750}{1785} \approx 0.420$. The approximate 95% confidence interval is calculated as follows.

$$\hat{p} \pm 2\sqrt{\frac{\hat{p}(1-\hat{p})}{n}} = 0.420 \pm 2\sqrt{\frac{0.420(1-0.420)}{1785}} = 0.420 \pm 2\sqrt{\frac{0.420(0.580)}{1785}} \approx 0.420 \pm 0.023$$

$$0.420 - 0.023 = 0.397 \text{ to } 0.420 + 0.023 = 0.443$$

45. (a) The sample proportion who admit running a red light is $\hat{p} = \frac{171}{880} \approx 0.194$. The approximate 95% confidence interval is calculated as follows.

$$\hat{p} \pm 2\sqrt{\frac{\hat{p}(1-\hat{p})}{n}} = 0.194 \pm 2\sqrt{\frac{0.194(1-0.194)}{880}} = 0.194 \pm 2\sqrt{\frac{0.194(0.806)}{880}} \approx 0.194 \pm 0.027$$

$$0.194 - 0.027 = 0.167 \text{ to } 0.194 + 0.027 = 0.221$$

(b) It is likely that more than 171 ran a red light, because some people are reluctant to admit illegal or antisocial acts.

47. (a) $\hat{p} = 0.5$

(b) $2\sqrt{\dfrac{\hat{p}(1-\hat{p})}{n}} = 2\sqrt{\dfrac{0.5(1-0.5)}{n}} = 2\sqrt{\dfrac{0.5(0.5)}{n}} = 2(0.5)\sqrt{\dfrac{1}{n}} = \sqrt{\dfrac{1}{n}} = \dfrac{1}{\sqrt{n}}$

49. (a) $\frac{1468}{13,000} \approx 0.113 = 11.3\%$.

(b) The response rate is so low that it is likely that those who responded differ from the population as a whole. That is, there is a bias that the margin of error does not include.

51. (a) No. The number of e-filed returns in all states is much larger than the sample size. When this is true, the margin of error depends only on the size of the sample, not on the size of the population.

(b) The sample sizes vary from 970 to 49,000, so the margins of error will also vary.

53. The margin of error for 90% confidence comes from the central 90% of a normal sampling distribution. We need not go as far out to cover 90% of the distribution as to cover 95%. So the margin of error for 90% confidence is smaller than for 95% confidence.

55. The sample proportion of successes is $\hat{p} = \frac{7}{97} \approx 0.072$. That is, there were 7.2% successes in the sample. The approximate 95% confidence interval is calculated as follows.

$$\hat{p} \pm 2\sqrt{\frac{\hat{p}(1-\hat{p})}{n}} = 0.072 \pm 2\sqrt{\frac{0.072(1-0.072)}{97}} = 0.072 \pm 2\sqrt{\frac{0.072(0.928)}{97}} \approx 0.072 \pm 0.052$$

$$0.072 - 0.052 = 0.020 \text{ to } 0.072 + 0.052 = 0.124$$

We are 95% confident that the true proportion of articles that discuss the success of blinding is between 0.020 and 0.124 (that is, 2.0% to 12.4%).

57. The distribution of the sample proportion $\hat{p}$ is approximately normal with mean $p = 0.1$ (that is, 10%) and standard deviation

$$\sqrt{\frac{p(1-p)}{n}} = \sqrt{\frac{0.1(1-0.1)}{97}} = \sqrt{\frac{0.1(0.9)}{97}} \approx 0.030$$

or 3%. Notice that 7% is one standard deviation below the mean. By the 68 part of the 68-95-99.7 rule, 68% of all samples will have between 7% and 13% that discuss blinding. Half the remaining 32% lie on either side. So 16% of samples will have fewer than 7% articles that discuss blinding. That is, the probability is about 0.16.

Chapter 8
Probability: The Mathematics of Chance

Exercise Solutions

1. **(a)** Results will vary, but the probability of a head is usually greater than 0.5 when spinning pennies. One possible explanation is the "bottle cap effect." The rim on a penny is slightly wider on the head side, so just as spinning bottle caps almost always fall with the open side up, pennies fall more often with the head side up. Additional results will vary.

    (b) Results will vary.

3. The first five lines contain 200 digits, of which 21 are zeros. The proportion of zeros is $\frac{21}{200} = 0.105$.

TABLE 7.1	Random	Digits					
101	19223	95034	05756	28713	96409	12531	42544	82853
102	73676	47150	99400	01927	27754	42648	82425	36290
103	45467	71709	77558	00095	32863	29485	82226	90056
104	52711	38889	93074	60227	40011	85848	48767	52573
105	95592	94007	69971	91481	60779	53791	17297	59335

5. **(a)** $S = \{0, 1, 2, 3, 4, 5, 6, 7, 8, 9, 10\}$.

    (b) $S = \{0, 10, 20, 30, 40, 50, 60, 70, 80, 90, 100\}$.

    (c) $S = \{\text{Yes, No}\}$.

7. **(a)** $S = \{$ HHHH, HHHM, HHMH, HMHH, MHHH, HHMM, HMMH, MMHH, HMHM, MHHM, MHMH, HMMM, MMMH, MHMM, MMHM, MMMM$\}$.

    (b) $S = \{0, 1, 2, 3, 4\}$.

9. **(a)** The given probabilities have sum 0.81, so the probability of any other topic is $1 - 0.81 = 0.19$.

    (b) The probability of adult or scam is $0.145 + 0.142 = 0.287$.

11. The three events of tossing a pair a dice are independent events. There are 6 ways to toss doubles out of the 36 ways to toss the dice. Thus, each time Laurie tosses the dice, the probability that she will obtain doubles is $\frac{6}{36} = \frac{1}{6}$. Because there are three independent events, the probability that she will toss doubles three times is $\left(\frac{1}{6}\right)\left(\frac{1}{6}\right)\left(\frac{1}{6}\right) = \frac{1}{216}$.

13. (a) Here is the probability histogram:

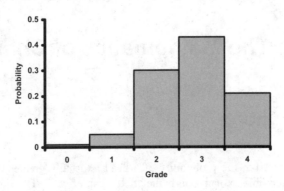

(b) $0.43 + 0.21 = 0.64.$

15. (a) No: the probabilities are between 0 and 1, but the sum is greater than 1. Rule 2 is violated.

$$0.56 + 0.24 + 0.44 + 0.17 = 1.41$$

(b) Yes: the probabilities are between 0 and 1, inclusively, and have sum 1.

$$0.39 + 0.28 + 0 + 0.33 = 1$$

One may find it surprising that no student surveyed was born in fall, but that outcome is possible.

17. Like in tossing two standard dice, the sample set contains a count between 2 and 12 with 36 possible outcomes. The probabilities of each count is the same as two standard dice.

sum		sum		sum		sum		sum		sum
2		4		5		6		7		9
3		5		6		7		8		10
3		5		6		7		8		10
4		6		7		8		9		11
4		6		7		8		9		11
5		7		8		9		10		12

Outcome	2	3	4	5	6	7	8	9	10	11	12
Probability	$\frac{1}{36}$	$\frac{1}{18}$	$\frac{1}{12}$	$\frac{1}{9}$	$\frac{5}{36}$	$\frac{1}{6}$	$\frac{5}{36}$	$\frac{1}{9}$	$\frac{1}{12}$	$\frac{1}{18}$	$\frac{1}{36}$

19. All 90 guests are equally likely to get the prize, so $P(\text{woman}) = \frac{42}{90} = \frac{7}{15}$.

21. (a) $2 \times 2 \times 2 \times 2 \times 2 \times 2 \times 2 \times 2 \times 2 \times 2 = 2^{10} = 1024$.

(b) $\frac{2}{1024} = \frac{1}{512}$.

23. (a) There are $36 \times 36 \times 36 = 36^3 = 46{,}656$ different codes. The probability of no x is as follows.

$$\frac{35 \times 35 \times 35}{46{,}656} = \frac{42{,}875}{46{,}656} \approx 0.919$$

(b) The probability of no digits is $\dfrac{26 \times 26 \times 26}{46{,}656} = \dfrac{17{,}576}{46{,}656} = \dfrac{2197}{5832} \approx 0.377$.

25. (a) The possibilities are *aps*, *asp*, *pas*, *psa*, *sap*, *spa*.

(b) "*asp*", "*pas*", "*sap*" and "*spa*" are English words.

(c) The probability is $\frac{4}{6} = \frac{2}{3} \approx 0.667$. The answer can also be expressed as exactly $66\frac{2}{3}\%$ or approximately 66.7%.

27. (a) There are 4 possible royal flush hands.

(b) There are $_{52}C_5 = \dfrac{52!}{5!(52-5)!} = \dfrac{52!}{5!47!} = \dfrac{52 \times 51 \times 50 \times 49 \times 48}{5 \times 4 \times 3 \times 2 \times 1} = 2{,}598{,}960$ possible 5-card hands.

(c) The probability would be $\dfrac{4}{2{,}598{,}960} = \dfrac{1}{649{,}740}$ or approximately 0.00000154.

29. (a) The area is $\frac{1}{2} \times \text{base} \times \text{height} = \frac{1}{2}(2)(1) = 1$.

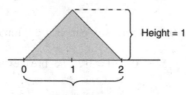

(b) Probability $\frac{1}{2}$ by symmetry or finding the area, $\frac{1}{2} \times \text{base} \times \text{height} = \frac{1}{2}(1)(1) = \frac{1}{2}$.

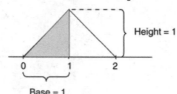

(c) The area representing this event is $\left(\frac{1}{2}\right)(0.5)(0.5) = 0.125$.

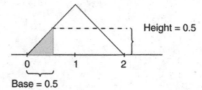

31. The probability would be $\frac{150}{600} = \frac{1}{4} = 0.25$.

33. The mean is as follows.

$$\mu = (0)(0.01)+(1)(0.05)+(2)(0.30)+(3)(0.43)+(4)(0.21)$$
$$= 0+0.05+0.60+1.29+0.84 = 2.78$$

The variance is as follows.

$$\sigma^2 = (0-2.78)^2(0.01)+(1-2.78)^2(0.05)+(2-2.78)^2(0.30)+(3-2.78)^2(0.43)+(4-2.78)^2(0.21)$$
$$= (-2.78)^2(0.01)+(-1.78)^2(0.05)+(-0.78)^2(0.30)+(0.22)^2(0.43)+(1.22)^2(0.21)$$
$$= (7.7284)(0.01)+(3.1684)(0.05)+(0.6084)(0.30)+(0.0484)(0.43)+(1.4884)(0.21)$$
$$= 0.077284+0.15842+0.18252+0.020812+0.312564$$
$$= 0.7516$$

Thus, the standard deviation is $\sigma = \sqrt{0.7516} \approx 0.8669$.

35. The mean for owner-occupied units is $\mu = (1)(0.000)+(2)(0.001)+...+(10)(0.047) = 6.248$.
For rented units, $\mu = (1)(0.011)+(2)(0.027)+...+(10)(0.005) = 4.321$.

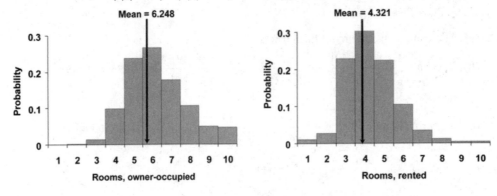

37. Both models have mean 1, because both density curves are symmetric about 1.

39. Since $\mu = (0)(0.85)+(200)(0.15) = 0+30 = 30$, the breakeven price would be $30 for the consumer.

41. A net prize would be what you have after expenses (cost to play, taxes, other fees, etc.). In this exercise, prizes are given as "net" prizes, so it is understood that they include the cost to play the game. The probability of receiving a prize is as follows.

$$\frac{1}{10,000,000} + \frac{1}{10,000} + \frac{1}{1,000} + \frac{3}{100}$$

$$= \frac{1}{10,000,000} + \frac{1000}{10,000,000} + \frac{10,000}{10,000,000} + \frac{300,000}{10,000,000} = \frac{311,001}{10,000,000}$$

Thus, the probability of not receiving a prize is $1 - \frac{311,001}{10,000,000} = \frac{9,688,999}{10,000,000}$.

We now have

$$\mu = (1,000,000)\left(\frac{1}{10,000,000}\right) + (1000)\frac{1}{10,000} + (100)\frac{1}{1000} + (4)\frac{3}{100} + (-1)\frac{9,688,999}{10,000,000}$$

$$= \frac{1}{10} + \frac{1}{10} + \frac{1}{10} + \frac{12}{100} + (-1)\frac{9,688,999}{10,000,000}$$

$$= \frac{3}{10} + \frac{12}{100} - \frac{9,688,999}{10,000,000}$$

$$= \frac{42}{100} - \frac{9,688,999}{10,000,000}$$

$$= \frac{4,200,000}{10,000,000} - \frac{9,688,999}{10,000,000}$$

$$= -\frac{5,488,999}{10,000,000} \approx -0.55$$

So approximately $0.45 "comes back" in prizes on the $1 ticket.

Another approach to this exercise is to make the prize a gross prize by adding back in the cost of play. By doing so, you need to only add four terms but the ease of calculation is reduced.

$$\mu = (1,000,001)\left(\frac{1}{10,000,000}\right) + (1001)\frac{1}{10,000} + (101)\frac{1}{1000} + (5)\frac{3}{100}$$

$$= \frac{1,000,001}{10,000,000} + \frac{1,001,000}{10,000,000} + \frac{1,010,000}{10,000,000} + \frac{1,500,000}{10,000,000}$$

$$= \frac{4,511,001}{10,000,000} \approx 0.45$$

Again, approximately $0.45 "comes back" in prizes on the $1 ticket.

43. (a) You would want $\mu = 0$ on any individual question. If we let x be the value of the question for getting it wrong by guessing, we have he following.

$$0 = (+1)\left(\tfrac{1}{5}\right) + x\left(\tfrac{4}{5}\right)$$
$$0 = \tfrac{1}{5} + \tfrac{4x}{5}$$
$$-\tfrac{1}{5} = \tfrac{4x}{5}$$
$$-1 = 4x$$
$$-\tfrac{1}{4} = x$$

A student should lose a quarter of a point for an incorrect answer.

(b) If you lose a quarter of a point for an incorrect answer, then we have the following possibilities.

Eliminate 1 wrong answer: $\mu = (+1)\left(\tfrac{1}{4}\right) + \left(-\tfrac{1}{4}\right)\left(\tfrac{3}{4}\right) = \tfrac{1}{16} = 0.0625$

Eliminate 2 wrong answers: $\mu = (+1)\left(\tfrac{1}{3}\right) + \left(-\tfrac{1}{4}\right)\left(\tfrac{2}{3}\right) = \tfrac{1}{6} \approx 0.1667$

Eliminate 3 wrong answers: $\mu = (+1)\left(\tfrac{1}{2}\right) + \left(-\tfrac{1}{4}\right)\left(\tfrac{1}{2}\right) = \tfrac{3}{8} = 0.375$

If you can eliminate any incorrect answers, there is an advantage to guessing.

45. Sample means $\bar{x}$ have a sampling distribution close to normal with mean $\mu = 0.15$ and standard deviation $\dfrac{\sigma}{\sqrt{n}} = \dfrac{0.4}{\sqrt{400}} = \dfrac{0.4}{20} = 0.02$. Therefore, 95% of all samples have an $\bar{x}$ between $0.15 - 2(0.02) = 0.15 - 0.04 = 0.11$ and $0.15 + 2(0.02) = 0.15 + 0.04 = 0.19$.

47. (a) The standard deviation of the average measurement is $\dfrac{\sigma}{\sqrt{n}} = \dfrac{10}{\sqrt{3}} \approx 5.77$ mg.

(b) To cut the standard deviation in half (from 10 mg to 5 mg), we need $n = 4$ measurements because $\dfrac{\sigma}{\sqrt{n}}$ is then $\dfrac{\sigma}{\sqrt{4}} = \dfrac{\sigma}{2}$. Averages of several measurements are less variable than individual measurements, so an average is more likely to give about the same result each time.

49. (a) Sketch a normal curve and mark the center at 4600 and the change-of-curvature points at 4590 and 4610. The curve will extend from about 4570 to 4630. This is the curve for one measurement. The mean of $n = 3$ measurements has mean $\mu = 4600$ mg and standard deviation 5.77 mg. Mark points about 5.77 above and below 4600 and sketch a second curve.

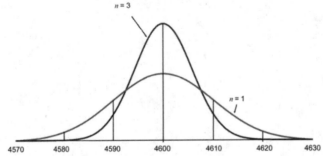

(b) Use the 95 part of the 68-95-99.7 rule with $\sigma = 10$.

$$4600 - 2(10) = 4600 - 20 = 4580 \text{ to } 4600 + 2(10) = 4600 + 20 = 4620$$

(c) Now the standard deviation is 5.77, so we have the following.

$$4600 - 2(5.77) = 4600 - 11.54 = 4588.46 \text{ to } 4600 + 2(5.77) = 4600 + 11.54 = 4611.54$$

51.

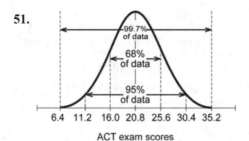

ACT exam scores

(a) Because 25.6 is one standard deviation above the mean, the probability is about 0.16.

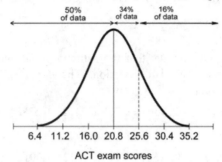

ACT exam scores

(b) The mean remains $\mu = 20.8$. The standard deviation is $\dfrac{\sigma}{\sqrt{9}} = \dfrac{4.8}{3} = 1.6$.

(c) Because $25.6 = 20.8 + 4.8 = 20.8 + 3(1.6)$ is three standard deviations above the mean, the probability is about 0.0015. (This is half of the 0.003 probability for outcomes more than three standard deviations from the mean, using the 99.7 part of the 68-95-99.7 rule.)

53. (a) There are $26 \times 10 \times 10 \times 26 \times 26 \times 26 = 45,697,600$ different license plates of this form.

(b) There are $26 \times 10 \times 10 = 2600$ plates ending in AAA, because that leaves only the first three characters free.

(c) The probability is $\dfrac{2600}{45,697,600} \approx 0.0000569$.

55. (a) The probability is $0.10 + 0.08 = 0.18$.

(b) The complement to the event of working out at least one day is working out no days. Thus, using the complement rule, the desired probability is $1 - 0.61 = 0.39$.

57. (a) The variance is as follows.

$$\sigma^2 = (0-0.77)^2(0.61)+(1-0.77)^2(0.17)+(2-0.77)^2(0.10)+(3-0.77)^2(0.08)+(4-0.77)^2(0.04)$$

$$= (-0.77)^2(0.61)+(0.23)^2(0.17)+(1.23)^2(0.10)+(2.23)^2(0.08)+(3.23)^2(0.04)$$

$$= (0.5929)(0.61)+(0.0529)(0.17)+(1.5129)(0.10)+(4.9729)(0.08)+(10.4329)(0.04)$$

$$= 0.361669+0.008993+0.15129+0.397832+0.417316$$

$$= 1.3371$$

Thus, the standard deviation is $\sigma = \sqrt{1.3371} \approx 1.156$ days.

(b) The mean, $\bar{x}$, of $n = 100$ observations has mean $\mu = 0.77$ and standard deviation

$$\frac{\sigma}{\sqrt{n}} = \frac{1.156}{\sqrt{100}} = \frac{1.156}{10} \approx 0.1156.$$

The Central Limit Theorem says that the sampling distribution of $\bar{x}$ is approximately normal with this mean and standard deviation. The 95 part of the 68-95-99.7 rule says that with probability 0.95, values of $\bar{x}$ lie between

$$0.77 - 2(0.1156) = 0.77 - 0.2312 = 0.5388 \approx 0.54 \text{ day}$$

and

$$0.77 + 2(0.1156) = 0.77 + 0.2312 = 1.0012 \approx 1.00 \text{ day}.$$

59. (a) Since there are only 7 days in the week and 10 people chosen at random, the probability that there is a match is 100% (certainty).

(b) The complement to the event of at least one match in the day of the month these 10 people were born is no match in the day of the month these 10 people were born. With the assumption of 31 days in a month, the probability we seek is as follows.

$$1-\left(\tfrac{30}{31}\right)\times\left(\tfrac{29}{31}\right)\times\cdots\times\left(\tfrac{22}{31}\right) = 1-\tfrac{31P_{10}}{31^{10}} \approx 1-0.196 = 0.804 \text{ or approximately } 80\%$$

(c) The complement to the event of at least one match in the day of the year these 10 people were born is no match in the day of the year these 10 people were born. With the assumption of 365 days in a year, the probability we seek is as follows.

$$1-\left(\tfrac{364}{365}\right)\times\left(\tfrac{363}{365}\right)\times\cdots\times\left(\tfrac{355}{365}\right) = 1-\tfrac{365P_{10}}{365^{10}} \approx 1-0.883 = 0.117 \text{ or approximately } 12\%$$

Chapter 9
Social Choice: The Impossible Dream

Exercise Solutions

1. Minority Rule satisfies condition (1): An exchange of marked ballots between two voters leaves the number of votes for each candidate unchanged, so whichever candidate won on the basis of having fewer votes before the exchange still has fewer votes after the exchange. Minority rule also satisfies condition (2): Suppose candidate X receives n votes and candidate Y receives m votes, and candidate X wins because $n < m$. Now suppose that a new election is held, and every voter reverses his or her vote. Then candidate X has m votes and candidate Y has n votes, and so candidate Y is the winner. Minority rule, however, fails condition (3): Suppose, for example, that there are 3 voters and that candidate X wins 1 out of the 3 votes. Now suppose that one of the 2 voters who voted for candidate Y reverses his or her vote. Then candidate X would have 2 votes, and candidate Y would have 1 vote, thus resulting in a win for candidate Y.

3. A dictatorship satisfies condition (2): If a new election is held and every voter (in particular, the dictator) reverses his or her ballot, then certainly the outcome of the election is reversed. A dictatorship also satisfies condition (3): If a single voter changes his or her ballot from being a vote for the loser of the previous election to a vote for the winner of the previous election, then the single voter could not have been a dictator (since the dictator's ballot was not a vote for the loser of the previous election). Thus, the outcome of the new election is the same as the outcome of the previous election. A dictatorship, however, fails to satisfy condition (1): If the dictator exchanges his/her marked ballot with any voter whose marked ballot differs from that of the dictator, then the outcome of the election is certainly reversed.

5. With an odd number of voters, each one-on-one score will have a winner because there cannot be a tie. If one of the candidates, say A, is the winner by Condorcet's method, then A would have beaten every other candidate in a one-on-one competition. It would be impossible for another candidate, say B, to have beaten A. Hence, B could not also be a winner.

7. With the following four preference lists, we can see that three-fourths prefer A to B, three-fourths prefer B to C, three-fourths prefer C to D, and three-fourths prefer D to A.

Rank	Number of voters (4)			
	1	1	1	1
First	A	B	C	D
Second	B	C	D	A
Third	C	D	A	B
Fourth	D	A	B	C

9.

22%	23%	15%	29%	7%	4%
D	D	H	H	J	J
H	J	D	J	H	D
J	H	J	D	D	H

(a) We must check the one-on-one scores of D versus H, D versus J, and H versus J.

D versus H:D is over H on 22% + 23% + 4% = 49% of the ballots, while the reverse is true on 15% + 29% + 7% = 51%. Thus, H defeats D, 51% to 49%.

D versus J: D is over J on 22% + 23% + 15% = 60% of the ballots, while the reverse is true on 29% + 7% + 4% = 40%. Thus, D defeats J, 60% to 40%.

H versus J: H is over J on 22% + 15% + 29% = 66% of the ballots, while the reverse is true on 23% + 7% + 4% = 34%. Thus, H defeats J, 66% to 34%.

Yes, there is a Condorcet winner. Since H can defeat both D and J in a one-to one competition, Elizabeth Holtzman (H) is the winner by Condorcet's method.

(b) D has 22% + 23% = 45% of the first-place votes. H has 15% + 29% = 44% of the first-place votes. J has 7% + 4% = 11% of the first-place votes. Since D has the most first-place votes, Alfonse D'Amato (D) is the winner by plurality voting.

11.

	Number of voters (7)				
Rank	**2**	**2**	**1**	**1**	**1**
First	C	D	C	B	A
Second	A	A	D	D	D
Third	B	C	A	A	B
Fourth	D	B	B	C	C

(a) A has 1 first-place vote. B has 1 first-place vote. C has 2 + 1 = 3 first-place votes. D has 2 first-place votes. Since C has the most first-place votes, C is the winner by plurality voting.

(b)

Preference	1st place votes × 3	2nd place votes × 2	3rd place votes × 1	4th place votes × 0	Borda score
A	1 × 3	4 × 2	2 × 1	0 × 0	13
B	1 × 3	0 × 2	3 × 1	3 × 0	6
C	3 × 3	0 × 2	2 × 1	2 × 0	11
D	2 × 3	3 × 2	0 × 1	2 × 0	12

Thus, A has the highest Borda score and is declared the winner.

Note, the Borda score for each candidate could also have been determined firstly by individually replacing the candidates below the one you are determining the score for by a box.

For A:

	Number of voters (7)				
Rank	**2**	**2**	**1**	**1**	**1**
First	C	D	C	B	A
Second	A	A	D	D	□
Third	□	□	A	A	□
Fourth	□	□	□	□	□

To show that the Borda score for candidate A is 13, it needs to be noted that each box below A counts 2 times in the first and second columns. The Borda score for A is therefore $(4 \times 2) + (5 \times 1) = 8 + 5 = 13$.

Continued on next page

11. (b) continued

For B:

	Number of voters (7)				
Rank	**2**	**2**	**1**	**1**	**1**
First	C	D	C	B	A
Second	A	A	D	☐	D
Third	B	C	A	☐	B
Fourth	☐	B	B	☐	☐

To show that the Borda score for candidate B is 6, it needs to be noted that each box below B counts 2 times in the first column. The Borda score for B is $(1 \times 2) + (4 \times 1) = 2 + 4 = 6$.

For C:

	Number of voters (7)				
Rank	**2**	**2**	**1**	**1**	**1**
First	C	D	C	B	A
Second	☐	A	☐	D	D
Third	☐	C	☐	A	B
Fourth	☐	☐	☐	C	C

To show that the Borda score for candidate C is 11, it needs to be noted that each box below C counts 2 times in the first and second columns. The Borda score for C is therefore $(4 \times 2) + (3 \times 1) = 8 + 3 = 11$.

For D:

	Number of voters (7)				
Rank	**2**	**2**	**1**	**1**	**1**
First	C	D	C	B	A
Second	A	☐	D	D	D
Third	B	☐	☐	☐	☐
Fourth	D	☐	☐	☐	☐

To show that the Borda score for candidate D is 12, it needs to be noted that each box below D counts 2 times in the second column. The Borda score for D is $(3 \times 2) + (6 \times 1) = 6 + 6 = 12$.

(c) Since A and B have the least number of first-place votes (see Part a), both are eliminated.

	Number of voters (7)				
Rank	**2**	**2**	**1**	**1**	**1**
First	C	D	C	D	D
Second	D	C	D	C	C

C now has $2 + 1 = 3$ first-place votes. D now has $2 + 1 + 1 = 4$ first-place votes. Thus, D is the winner by the Hare system.

(d) In sequential pairwise voting with the agenda $B, D, C, A,$ we first pit B against D. There are 3 voters who prefer B to D and 4 prefer D to B. Thus, D wins by a score of 4 to 3. B is therefore eliminated, and D moves on to confront C.

There are 4 voters who prefer D to C and 3 prefer C to D. Thus, D wins by a score of 4 to 3. C is therefore eliminated, and D moves on to confront A.

There are 4 voters who prefer D to A and 3 prefer A to D. Thus, D wins by a score of 4 to 3.

Thus, D is the winner by sequential pairwise voting with the agenda B, D, C, A.

13.

	Number of voters (5)				
Rank	**1**	**1**	**1**	**1**	**1**
First	A	B	C	D	E
Second	B	C	B	C	D
Third	E	A	E	A	C
Fourth	D	D	D	E	A
Fifth	C	E	A	B	B

(a) A has 1 first-place vote. B has 1 first-place vote. C has 1 first-place vote. D has 1 first-place vote. E has 1 first-place vote. Since all candidates have the same number of first-place votes, they tie.

(b)

Preference	1st place votes × 4	2nd place votes × 3	3rd place votes × 2	4th place votes × 1	5th place votes × 0	Borda score
A	1×4	0×3	2×2	1×1	1×0	9
B	1×4	2×3	0×2	0×1	2×0	10
C	1×4	2×3	1×2	0×1	1×0	12
D	1×4	1×3	0×2	3×1	0×0	10
E	1×4	0×3	2×2	1×1	1×0	9

Thus, C has the highest Borda score and is declared the winner.

Note, the Borda score for each candidate could also have been determined firstly by individually replacing the candidates below the one you are determining the score for by a box.

For A:

	Number of voters (5)				
Rank	**1**	**1**	**1**	**1**	**1**
First	A	B	C	D	E
Second	☐	C	B	C	D
Third	☐	A	E	A	C
Fourth	☐	☐	D	☐	A
Fifth	☐	☐	A	☐	☐

The Borda score for candidate A is 9 since there are 9 boxes.

For B:

	Number of voters (5)				
Rank	**1**	**1**	**1**	**1**	**1**
First	A	B	C	D	E
Second	B	☐	B	C	D
Third	☐	☐	☐	A	C
Fourth	☐	☐	☐	E	A
Fifth	☐	☐	☐	B	B

The Borda score for candidate B is 10 since there are 10 boxes.

For C:

	Number of voters (5)				
Rank	**1**	**1**	**1**	**1**	**1**
First	A	B	C	D	E
Second	B	C	☐	C	D
Third	E	☐	☐	☐	C
Fourth	D	☐	☐	☐	☐
Fifth	C	☐	☐	☐	☐

The Borda score for candidate C is 12.

Continued on next page

13. (b) continued

For D:

	Number of voters (5)				
Rank	**1**	**1**	**1**	**1**	**1**
First	A	B	C	D	E
Second	B	C	B	□	D
Third	E	A	E	□	□
Fourth	D	D	D	□	□
Fifth	□	□	□	□	□

The Borda score for candidate D is 10 since there are 10 boxes.

For E:

	Number of voters (5)				
Rank	**1**	**1**	**1**	**1**	**1**
First	A	B	C	D	E
Second	B	C	B	C	□
Third	E	A	E	A	□
Fourth	□	D	□	E	□
Fifth	□	E	□	□	□

The Borda score for candidate E is 9 since there are 9 boxes.

(c) Since all candidates have the same least number of first-place votes (see Part a), they all tie.

(d) In sequential pairwise voting with the agenda A, B, C, D, E, we first pit A against B. There are 3 voters who prefer A to B and 2 prefer B to A. Thus, A wins by a score of 3 to 2. B is therefore eliminated, and A moves on to confront C.

There is 1 voter who prefers A to C and 4 prefer C to A. Thus, C wins by a score of 4 to 1. A is therefore eliminated, and C moves on to confront D.

There are 2 voters who prefer C to D and 3 prefer D to C. Thus, D wins by a score of 3 to 2. C is therefore eliminated, and D moves on to confront E.

There are 2 voters who prefer D to E and 3 prefer E to D. Thus, E wins by a score of 3 to 2. Thus, E is the winner by sequential pairwise voting with the agenda A, B, C, D, E.

15.

	Number of voters (7)				
Rank	**2**	**2**	**1**	**1**	**1**
First	C	E	C	D	A
Second	E	B	A	E	E
Third	D	D	D	A	C
Fourth	A	C	E	C	D
Fifth	B	A	B	B	B

(a) A has 1 first-place vote. B has 0 first-place votes. C has $2 + 1 = 3$ first-place votes. D has 1 first-place vote. E has 2 first-place votes. Since C has the most first-place votes, C is the winner by plurality voting.

(b)

Preference	1st place votes × 4	2nd place votes × 3	3rd place votes × 2	4th place votes × 1	5th place votes × 0	Borda score
A	1×4	1×3	1×2	2×1	2×0	11
B	0×4	2×3	0×2	0×1	5×0	6
C	3×4	0×3	1×2	3×1	0×0	17
D	1×4	0×3	5×2	1×1	0×0	15
E	2×4	4×3	0×2	1×1	0×0	21

Thus, E has the highest Borda score and is declared the winner.

Continued on next page

15. (b) continued

Note, the Borda score for each candidate could also have been determined firstly by individually replacing the candidates below the one you are determining the score for by a box.

For A:

	Number of voters (7)				
Rank	**2**	**2**	**1**	**1**	**1**
First	C	E	C	D	A
Second	E	B	A	E	☐
Third	D	D	☐	A	☐
Fourth	A	C	☐	☐	☐
Fifth	☐	A	☐	☐	☐

To show that the Borda score for candidate A is 11, it needs to be noted that each box below A counts 2 times in the first column. The Borda score for A is $(1 \times 2) + (9 \times 1) = 2 + 9 = 11$.

For B:

	Number of voters (7)				
Rank	**2**	**2**	**1**	**1**	**1**
First	C	E	C	D	A
Second	E	B	A	E	E
Third	D	☐	D	A	C
Fourth	A	☐	E	C	D
Fifth	B	☐	B	B	B

To show that the Borda score for candidate B is 6, it needs to be noted that each box below B counts 2 times in the second column. The Borda score for B is $3 \times 2 = 6$.

For C:

	Number of voters (7)				
Rank	**2**	**2**	**1**	**1**	**1**
First	C	E	C	D	A
Second	☐	B	☐	E	E
Third	☐	D	☐	A	C
Fourth	☐	C	☐	C	☐
Fifth	☐	☐	☐	☐	☐

To show that the Borda score for candidate C is 17, it needs to be noted that each box below C counts 2 times in the first and second columns. The Borda score for C is therefore $(5 \times 2) + (7 \times 1) = 10 + 7 = 17$.

For D:

	Number of voters (7)				
Rank	**2**	**2**	**1**	**1**	**1**
First	C	E	C	D	A
Second	E	B	A	☐	E
Third	D	D	D	☐	C
Fourth	☐	☐	☐	☐	D
Fifth	☐	☐	☐	☐	☐

To show that the Borda score for candidate D is 15, it needs to be noted that each box below D counts 2 times in the first and second columns. The Borda score for D is therefore $(4 \times 2) + (7 \times 1) = 8 + 7 = 15$.

Continued on next page

15. (b) continued

For E:

	Number of voters (7)				
Rank	**2**	**2**	**1**	**1**	**1**
First	C	E	C	D	A
Second	E	☐	A	E	E
Third	☐	☐	D	☐	☐
Fourth	☐	☐	E	☐	☐
Fifth	☐	☐	☐	☐	☐

To show that the Borda score for candidate E is 21, it needs to be noted that each box below E counts 2 times in the first and second columns. The Borda score for E is $(7 \times 2) + (7 \times 1) = 14 + 7 = 21$.

(c) Since B has the least number of first-place votes (see Part a), B is eliminated.

	Number of voters (7)				
Rank	**2**	**2**	**1**	**1**	**1**
First	C	E	C	D	A
Second	E	D	A	E	E
Third	D	C	D	A	C
Fourth	A	A	E	C	D

A now has 1 first-place vote. C now has $2 + 1 = 3$ first-place votes. D now has 1 first-place vote. E has 2 first-place votes. Since A and D have the least number of first-place votes, they are both eliminated.

	Number of voters (7)				
Rank	**2**	**2**	**1**	**1**	**1**
First	C	E	C	E	E
Second	E	C	E	C	C

C now has $2 + 1 = 3$ first-place votes. E now has $2 + 1 + 1 = 4$ first-place votes. Thus, E is the winner by the Hare system.

(d) In sequential pairwise voting with the agenda A, B, C, D, E, we first pit A against B. There are 5 voters who prefer A to B and 2 prefer B to A. Thus, A wins by a score of 5 to 2. B is therefore eliminated, and A moves on to confront C.

There are 2 voters who prefer A to C and 5 prefer C to A. Thus, C wins by a score of 5 to 2. A is therefore eliminated, and C moves on to confront D.

There are 4 voters who prefer C to D and 3 prefer D to C. Thus, C wins by a score of 4 to 3. D is therefore eliminated, and C moves on to confront E.

There are 3 voters who prefer C to E and 4 prefer E to C. Thus, E wins by a score of 4 to 3.

Thus, E is the winner by sequential pairwise voting with the agenda A, B, C, D, E.

17.

Rank	Number of voters (7)						
	1	1	1	1	1	1	1
First	C	D	C	B	E	D	C
Second	A	A	E	D	D	E	A
Third	E	E	D	A	A	A	E
Fourth	B	C	A	E	C	B	B
Fifth	D	B	B	C	B	C	D

(a) A has 0 last-place votes. B has 3 last-place votes. C has 2 last-place votes. D has 2 last-place votes. E has 0 last-place votes. Since B has the most last-place votes, B is eliminated.

Rank	Number of voters (7)						
	1	1	1	1	1	1	1
First	C	D	C	D	E	D	C
Second	A	A	E	A	D	E	A
Third	E	E	D	E	A	A	E
Fourth	D	C	A	C	C	C	D

A has 1 last-place vote. C has 4 last-place votes. D has 2 last-place votes. E has 0 last-place votes. Since C has the most last-place votes, C is eliminated.

Rank	Number of voters (7)						
	1	1	1	1	1	1	1
First	A	D	E	D	E	D	A
Second	E	A	D	A	D	E	E
Third	D	E	A	E	A	A	D

A has 3 last-place votes. D has 2 last-place votes. E has 2 last-place votes. Since A has the most number of the last-place votes, A is eliminated.

Rank	Number of voters (7)						
	1	1	1	1	1	1	1
First	E	D	E	D	E	D	E
Second	D	E	D	E	D	E	D

D now has 4 last-place votes. E now has 3 last-place votes. Thus, E is the winner by the procedure of Clyde Coombs.

(b) To show that it is possible for two voters and three candidates to result in different outcomes using different methods (Coombs procedure and the Hare method), we need to find an example that illustrates such an occurrence.

One possible scenario is having candidates A, B, and C with the following preference lists.

Rank	Number of voters (2)	
	1	1
First	A	C
Second	B	B
Third	C	A

Using the Coombs procedure, both C and A have the same number of last-place votes. They are both therefore eliminated leaving only B. B therefore is the winner using the Coombs procedure.

Using the Hare method, B has the least number of first-place votes. B is therefore eliminated resulting in the following preference list.

Rank	Number of voters (2)	
	1	1
First	A	C
Second	C	A

Since A and C both have the same number of first-place votes, a tie is declared.

19. (a) Plurality voting satisfies the Pareto condition because if everyone prefers B to D, for example, then D has no first-place votes at all. Thus, D cannot be among the winners in plurality voting.

(b) Plurality voting satisfies the monotonicity because if a candidate wins on the basis of having the most first-place votes, then moving that candidate up one spot on some list (and making no other changes) neither decreases the number of first-place votes for the winning candidate nor increases the number of first-place votes for any other candidate. Hence, the original winner remains a winner in plurality voting.

21. (a) Sequential pairwise voting satisfies the Condorcet winner criterion because a Condorcet winner always wins the kind of one-on-one contest that is used to produce the winner in sequential pairwise voting.

(b) Sequential pairwise voting satisfies the monotonicity because moving a candidate up on some list only improves that candidate's chances in one-on-one contests.

23. In the plurality runoff method, in order to have one candidate's ranking be consistently higher than another candidate's would imply that only one candidate would be considered. This candidate would have received all first-place votes and is therefore the winner. Thus, none of the other candidates are being considered and cannot be the winner. Pareto condition is therefore satisfied.

25.

	Number of voters (13)				
Rank	**4**	**3**	**3**	**2**	**1**
First	A	B	C	D	E
Second	B	A	A	B	D
Third	C	C	B	C	C
Fourth	D	D	D	A	B
Fifth	E	E	E	E	A

Since A has the highest number of first-place votes and B and C have the same number of second-place votes, D and E are eliminated.

	Number of voters (13)				
Rank	**4**	**3**	**3**	**2**	**1**
First	A	B	C	B	C
Second	B	A	A	C	B
Third	C	C	B	A	A

A now has 4 first-place votes. B has $3 + 2 = 5$ first-place votes. C has $3 + 1 = 4$ first-place votes. Since B has the most first-place votes, B is the winner by the plurality runoff method.

Now if the single voter on the far right in our original preference lists moved B to the top, we would have the following.

	Number of voters (13)				
Rank	**4**	**3**	**3**	**2**	**1**
First	A	B	C	D	B
Second	B	A	A	B	E
Third	C	C	B	C	D
Fourth	D	D	D	A	C
Fifth	E	E	E	E	A

Since A and B have the same number of first-place votes, C, D and E are eliminated.

	Number of voters (13)				
Rank	**4**	**3**	**3**	**2**	**1**
First	A	B	A	B	B
Second	B	A	B	A	A

A now has $4 + 3 = 7$ first-place votes. B now has $3 + 2 + 1 = 6$ first-place votes. Since A has the majority of the first-place votes, A is the winner by the plurality runoff method. This does not satisfy monotonicity since a ballot change favorable to B resulted in B losing.

27. One possible scenario is having candidates A, B, and C with the following preference lists.

Rank	Number of voters (5)	
	3	2
First	A	B
Second	B	C
Third	C	A

Preference	1st place votes $\times$ 2	2nd place votes $\times$ 1	3rd place votes $\times$ 0	Borda score
A	3×2	0×1	2×0	6
B	2×2	3×1	0×0	7
C	0×2	2×1	3×0	2

Thus, B has the highest Borda score and is declared the winner.

Note, the Borda score for each candidate could also have been determined firstly by individually replacing the candidates below the one you are determining the score for by a box.

For A:

Rank	Number of voters (5)	
	3	2
First	A	B
Second	$\square$	C
Third	$\square$	A

To show that the Borda score for candidate A is 6, it needs to be noted that each box below A counts 3 times in the first column. The Borda score for A is $3 \times 2 = 6$.

For B:

Rank	Number of voters (5)	
	3	2
First	A	B
Second	B	$\square$
Third	$\square$	$\square$

To show that the Borda score for candidate B is 7, it needs to be noted that each box below B counts 3 times in the first column and 2 times in the second column. The Borda score for B is $(1 \times 3) + (2 \times 2) = 3 + 4 = 7$.

For C:

Rank	Number of voters (5)	
	3	2
First	A	B
Second	B	C
Third	C	$\square$

To show that the Borda score for candidate C is 2, it needs to be noted that each box below C counts 2 times in the second columns. The Borda score for C is $1 \times 2 = 2$.

Now using Condorcet's method, we must check the one-on-one scores of A versus B, A versus C, and B versus C.

    A versus B: A is over B on 3 ballots, while the reverse is true on 2 ballots. Thus, A defeats B, 3 to 2.

    A versus C: A is over C on 3 ballots, while the reverse is true on 2 ballots. Thus, A defeats C, 3 to 2.

    B versus C: B is over C on all 5 ballots. Thus, B defeats C, 5 to 0.

The winner by Condorcet's method is A.

In this case the Borda count produces B as the winner while A is the Condorcet winner. Thus, this example shows that the Borda count does not satisfy the Condorcet winner criterion.

29.

	Number of voters (21)			
Rank	7	6	5	3
First	A	B	C	D
Second	B	A	B	C
Third	C	C	A	B
Fourth	D	D	D	A

(a) Since *D* has the least number of first-place votes, *D* is eliminated.

	Number of voters (21)			
Rank	7	6	5	3
First	A	B	C	C
Second	B	A	B	B
Third	C	C	A	A

A now has 7 first-place votes. *B* now has 6 first-place votes. *C* now has 5 + 3 = 8 first-place votes. Since *B* has the least number of first-place votes, *B* is eliminated.

	Number of voters (21)			
Rank	7	6	5	3
First	A	A	C	C
Second	C	C	A	A

A now has 7 + 6 = 13 first-place votes. *C* now has 5 + 3 = 8 first-place votes. Thus, *A* is the unique winner by the Hare system.

(b)

	Number of voters (21)			
Rank	7	6	5	3
First	A	B	C	A
Second	B	A	B	D
Third	C	C	A	C
Fourth	D	D	D	B

Since *D* has the least number of first-place votes, *D* is eliminated.

	Number of voters (21)			
Rank	7	6	5	3
First	A	B	C	A
Second	B	A	B	C
Third	C	C	A	B

A now has 7 + 3 =10 first-place votes. *B* now has 6 first-place votes. *C* now has 5 first-place votes. Since *C* has the least number of first-place votes, *C* is eliminated.

	Number of voters (21)			
Rank	7	6	5	3
First	A	B	B	A
Second	B	A	A	B

A now has 7 + 3 = 10 first-place votes. *B* now has 6 + 5 = 11 first-place votes. Thus, *B* is the winner by the Hare system.

31. If a candidate, say B, is ranked last on a majority of votes (over 50% of the votes are for last-place) then this candidate may or may not be considered in the runoff. Obviously, if this candidate is not considered in the runoff, then this candidate cannot be among the winners.

Now suppose this candidate is considered in the runoff. Since we assume there are no ties for second-place, then candidate B is having a runoff against another candidate, say candidate A. Since candidate B has the majority of the last-place votes, candidate A must have the majority of the first place votes and is hence the winner.

33. (a) If each alternative has exactly one first-place vote, then the outcome of the election is a three-way tie using the Hare procedure. In plurality runoff, the candidates with the most first-place votes have a runoff. Since they all have the same number of first-place votes, a runoff will produce the same results, which is a three-way tie.

(b) In both the Hare procedure and plurality runoff, the candidate with two or more first-place votes will become the winner.

(c) No; either the situation in part (a) or the situation in part (b) must occur.

35.

					Voters					
Candidates	**1**	**2**	**3**	**4**	**5**	**6**	**7**	**8**	**9**	**10**
A	X	X	X			X	X	X		X
B		X	X	X	X	X	X	X	X	
C			X					X		
D	X	X	X	X	X		X	X	X	X
E	X		X		X		X		X	
F	X		X	X	X	X	X	X		X
G	X	X	X	X	X			X		
H		X		X		X		X		X

A has 7 approval votes, B has 8, C has 2, D has 9, E has 5, F has 8, G has 6, and H has 5. Ranking the candidates we have, D (9), B and F (8), A (7), G (6), E and H (5), and C (2).

(a) Since D has the most votes, D is chosen for the board.

(b) The top four are A, B, D, and F.

(c) Candidates B, D and F have at least 80% (8 out of 10) approval.

(d) Candidates A, B, D, F and G have at least 60% (6 out of 10) approval. But since at most four candidates can be elected, only A, B, D, and F are considered.

Chapter 10
The Manipulability of Voting Systems

Exercise Solutions

1. One example of two such elections is the following:

Election 1

Rank	Number of voters (3)		
First	A	A	B
Second	B	B	A

Election 2

Rank	Number of voters (3)		
First	B	A	B
Second	A	B	A

With the voting system in which the candidate with the fewest first-place votes wins, B is the winner in the first election. However, if the leftmost voter changes his or her ballot as shown in the second election, then A becomes the winner. Taking the ballots in the first election to be the sincere preferences of the voters, the leftmost voter (who prefers A to B) has secured a more favorable outcome by the submission of a disingenuous ballot.

3. One example of two such elections is the following:

Election 1

Rank	Number of voters (3)		
First	A	B	B
Second	B	A	A

Election 2

Rank	Number of voters (3)		
First	B	B	B
Second	A	A	A

With the voting system in which the candidate receiving an even number of first-place votes wins, B is the winner in the first election. However, if the leftmost voter changes his or her ballot as shown in the second election, then A becomes the winner. Taking the ballots in the first election to be the sincere preferences of the voters, the leftmost voter (who prefers A to B) has secured a more favorable outcome by the submission of a disingenuous ballot.

5. **(a)** The voting system does not treat all *voters* the same.

   (b) A dictatorship in which Voter #1 is the dictator.

   (c) A dictatorship in which Voter #2 is the dictator and a dictatorship in which Voter #3 is the dictator.

7.   Election 1

	Number of voters (2)	
Rank	**1**	**1**
First	B	A
Second	C	D
Third	A	C
Fourth	D	B

Preference	1st place votes × 3	2nd place votes × 2	3rd place votes × 1	4th place votes × 0	Borda score
A	1 × 3	0 × 2	1 × 1	0 × 0	4
B	1 × 3	0 × 2	0 × 1	1 × 0	3
C	0 × 3	1 × 2	1 × 1	0 × 0	3
D	0 × 3	1 × 2	0 × 1	1 × 0	2

With the given ballots, the winner using the Borda count is A. However, if the leftmost voter changes his or her preference ballot, we have the following.

Election 2

	Number of voters (2)	
Rank	**1**	**1**
First	C	A
Second	B	D
Third	D	C
Fourth	A	B

Preference	1st place votes × 3	2nd place votes × 2	3rd place votes × 1	4th place votes × 0	Borda score
A	1 × 3	0 × 2	0 × 1	1 × 0	3
B	0 × 3	1 × 2	0 × 1	1 × 0	2
C	1 × 3	0 × 2	1 × 1	0 × 0	4
D	0 × 3	1 × 2	1 × 1	0 × 0	3

With the new ballots, the winner using the Borda count is C.

9. Election 1

	Number of voters (3)		
Rank	**1**	**1**	**1**
First	*A*	*B*	*B*
Second	*B*	*A*	*A*
Third	*C*	*C*	*C*
Fourth	*D*	*D*	*D*

With the given ballots, the winner using the Borda count is *B*.

Preference	1st place votes × 3	2nd place votes × 2	3rd place votes × 1	4th place votes × 0	Borda score
A	1 × 3	2 × 2	0 × 1	0 × 0	7
B	2 × 3	1 × 2	0 × 1	0 × 0	8
C	0 × 3	0 × 2	3 × 1	0 × 0	3
D	0 × 3	0 × 2	0 × 1	3 × 0	0

The voter on the far left prefers *A* to *B*. By casting a disingenuous ballot (still preferring *A* to *B* though), the outcome of the election is altered.

Election 2

	Number of voters (3)		
Rank	**1**	**1**	**1**
First	*A*	*B*	*B*
Second	*C*	*A*	*A*
Third	*D*	*C*	*C*
Fourth	*B*	*D*	*D*

Preference	1st place votes × 3	2nd place votes × 2	3rd place votes × 1	4th place votes × 0	Borda score
A	1 × 3	2 × 2	0 × 1	0 × 0	7
B	2 × 3	0 × 2	0 × 1	1 × 0	6
C	0 × 3	1 × 2	2 × 1	0 × 0	4
D	0 × 3	0 × 2	1 × 1	2 × 0	1

Thus, *A* has the highest Borda score and is declared the winner.

11. The following is one such example:
Election 1

	Number of voters (9)								
Rank	1	1	1	1	1	1	1	1	1
First	A	B	B	A	D	A	F	A	F
Second	B	A	A	B	C	B	E	B	E
Third	C	C	C	C	B	C	D	C	D
Fourth	D	D	D	D	A	D	C	D	C
Fifth	E	E	E	E	E	E	B	E	B
Sixth	F	F	F	F	F	F	A	F	A

Preference	1st place votes × 5	2nd place votes × 4	3rd place votes × 3	4th place votes × 2	5th place votes × 1	6th place votes × 0	Borda score
A	4 × 5	2 × 4	0 × 3	1 × 2	0 × 1	2 × 0	30
B	2 × 5	4 × 4	1 × 3	0 × 2	2 × 1	0 × 0	31
C	0 × 5	1 × 4	6 × 3	2 × 2	0 × 1	0 × 0	26
D	1 × 5	0 × 4	2 × 3	6 × 2	0 × 1	0 × 0	23
E	0 × 5	2 × 4	0 × 3	0 × 2	7 × 1	0 × 0	15
F	2 × 5	0 × 4	0 × 3	0 × 2	0 × 1	7 × 0	10

Thus, B has the highest Borda score and is declared the winner. This was the expected result.

The voter on the far left prefers A to B. By casting a disingenuous ballot (still preferring A to B though), the outcome of the election is altered.

Election 2

	Number of voters (9)								
Rank	1	1	1	1	1	1	1	1	1
First	A	B	B	A	D	A	F	A	F
Second	D	A	A	B	C	B	E	B	E
Third	C	C	C	C	B	C	D	C	D
Fourth	B	D	D	D	A	D	C	D	C
Fifth	E	E	E	E	E	E	B	E	B
Sixth	F	F	F	F	F	F	A	F	A

Preference	1st place votes × 5	2nd place votes × 4	3rd place votes × 3	4th place votes × 2	5th place votes × 1	6th place votes × 0	Borda score
A	4 × 5	2 × 4	0 × 3	1 × 2	0 × 1	2 × 0	30
B	2 × 5	3 × 4	1 × 3	1 × 2	2 × 1	0 × 0	29
C	0 × 5	1 × 4	6 × 3	2 × 2	0 × 1	0 × 0	26
D	1 × 5	1 × 4	2 × 3	5 × 2	0 × 1	0 × 0	25
E	0 × 5	2 × 4	0 × 3	0 × 2	7 × 1	0 × 0	15
F	2 × 5	0 × 4	0 × 3	0 × 2	0 × 1	7 × 0	10

Thus, A has the highest Borda score and is declared the winner.

13. Election 1

	Number of voters (4)			
Rank	**1**	**1**	**1**	**1**
First	A	C	B	D
Second	B	A	D	C
Third	C	B	C	A
Fourth	D	D	A	B

Preference	1st place votes × 3	2nd place votes × 2	3rd place votes × 1	4th place votes × 0	Borda score
A	1 × 3	1 × 2	1 × 1	1 × 0	6
B	1 × 3	1 × 2	1 × 1	1 × 0	6
C	1 × 3	1 × 2	2 × 1	0 × 0	7
D	1 × 3	1 × 2	0 × 1	2 × 0	5

Thus, C has the highest Borda score and is declared the winner. But the winner becomes B if the leftmost voter changes his or her ballot as follows.

Election 2

	Number of voters (4)			
Rank	**1**	**1**	**1**	**1**
First	B	C	B	D
Second	A	A	D	C
Third	D	B	C	A
Fourth	C	D	A	B

Preference	1st place votes × 3	2nd place votes × 2	3rd place votes × 1	4th place votes × 0	Borda score
A	0 × 3	2 × 2	1 × 1	1 × 0	5
B	2 × 3	0 × 2	1 × 1	1 × 0	7
C	1 × 3	1 × 2	1 × 1	1 × 0	6
D	1 × 3	1 × 2	1 × 1	1 × 0	6

Thus, B has the highest Borda score and is declared the winner.

15. Election 1

	Number of voters (5)				
Rank	**1**	**1**	**1**	**1**	**1**
First	*A*	*B*	*B*	*A*	*A*
Second	*B*	*C*	*C*	*C*	*C*
Third	*C*	*A*	*A*	*B*	*B*

Since Candidates *A* and *B* both have the same (high) number of last-place votes, they are both eliminated, leaving Candidate *C* as the winner using Coombs rule. But the winner becomes *A* if the leftmost voter changes his or her ballot as the following shows.

Election 2

	Number of voters (5)				
Rank	**1**	**1**	**1**	**1**	**1**
First	*A*	*B*	*B*	*A*	*A*
Second	*C*	*C*	*C*	*C*	*C*
Third	*B*	*A*	*A*	*B*	*B*

B has the most last-place votes, thus Candidate *B* is eliminated, and we have the following.

	Number of voters (5)				
Rank	**1**	**1**	**1**	**1**	**1**
First	*A*	*C*	*C*	*A*	*A*
Second	*C*	*A*	*A*	*C*	*C*

C now has the most last-place votes, thus Candidate *C* is eliminated, and *A* becomes the winner by the Coombs method.

17. Election 1

Rank	Number of voters (5)				
	1	1	1	1	1
First	A	A	C	C	B
Second	B	B	A	A	C
Third	C	C	B	B	A

Since *A* and *C* have the most number of first-place votes, *B* is eliminated.

Rank	Number of voters (5)				
	1	1	1	1	1
First	A	A	C	C	C
Second	C	C	A	A	A

Since *C* has the most number of first-place votes, the winner using the plurality runoff rule is *C*. But the winner becomes *B* if the leftmost voter changes his or her ballot as the following shows.

Election 2

Rank	Number of voters (5)				
	1	1	1	1	1
First	B	A	C	C	B
Second	A	B	A	A	C
Third	C	C	B	B	A

Since *B* and *C* have the most number of first-place votes, *A* is eliminated.

Rank	Number of voters (5)				
	1	1	1	1	1
First	B	B	C	C	B
Second	C	C	B	B	C

Since *B* has the most number of first-place votes, the winner using the plurality runoff rule is *B*.

19.

	Number of voters (3)		
Rank	**1**	**1**	**1**
First	*A*	*C*	*B*
Second	*B*	*A*	*D*
Third	*D*	*B*	*C*
Fourth	*C*	*D*	*A*

(a) For *B* to win, consider the agenda *D*, *A*, *C*, *B*.

In sequential pairwise voting with the agenda *D*, *A*, *C*, *B*, we first pit *D* against *A*. There is 1 voter that prefers *D* to *A* and 2 prefer *A* to *D*. Thus, *A* wins by a score of 2 to 1. *D* is therefore eliminated, and *A* moves on to confront *C*.

There is 1 voter who prefers *A* to *C* and 2 prefer *C* to *A*. Thus, *C* wins by a score of 2 to 1. *A* is therefore eliminated, and *C* moves on to confront *B*.

There is 1 voter who prefers *C* to *B* and 2 prefer *B* to *C*. Thus, *B* wins by a score of 2 to 1.

Thus, *B* is the winner by sequential pairwise voting with the agenda *D*, *A*, *C*, *B*.

(b) For *C* to win, consider the agenda *B*, *D*, *A*, *C*.

In sequential pairwise voting with the agenda *B*, *D*, *A*, *C*, we first pit *B* against *D*. There are 3 voters that prefer *B* to *D* and 0 prefer *D* to *B*. Thus, *B* wins by a score of 3 to 0. *D* is therefore eliminated, and *B* moves on to confront *A*.

There is 1 voter who prefers *B* to *A* and 2 prefer *A* to *B*. Thus, *A* wins by a score of 2 to 1. *B* is therefore eliminated, and *A* moves on to confront *C*.

There is 1 voter who prefers *A* to *C* and 2 prefer *C* to *A*. Thus, *C* wins by a score of 2 to 1.

Thus, *C* is the winner by sequential pairwise voting with the agenda *B*, *D*, *A*, *C*.

(c) For *D* to win, consider the agenda *B*, *A*, *C*, *D*.

In sequential pairwise voting with the agenda *B*, *A*, *C*, *D*, we first pit *B* against *A*. There is 1 voter that prefers *B* to *A* and 2 prefer *A* to *B*. Thus, *A* wins by a score of 2 to 1. *B* is therefore eliminated, and *A* moves on to confront *C*.

There is 1 voter who prefers *A* to *C* and 2 prefer *C* to *A*. Thus, *C* wins by a score of 2 to 1. *A* is therefore eliminated, and *C* moves on to confront *D*.

There is 1 voter who prefers *C* to *D* and 2 prefer *D* to *C*. Thus, *D* wins by a score of 2 to 1.

Thus, *D* is the winner by sequential pairwise voting with the agenda *B*, *A*, *C*, *D*.

Note: In any of the three parts, the first two candidates can be switched and the outcome will be the same.

21. Election 1

22%	23%	15%	29%	7%	4%
D	D	H	H	J	J
H	J	D	J	H	D
J	H	J	D	D	H

D has 22% + 23% = 45% of the first-place votes. H has 15% + 29% = 44% of the first-place votes. J has 7% + 4% = 11% of the first-place votes. Since D has the most first-place votes, Alfonse D'Amato (D) is the winner by plurality voting. The plurality rule is group manipulable as the following shows if the voters in the 7% group all change their ballots.

Election 2

22%	23%	15%	29%	7%	4%
D	D	H	H	H	J
H	J	D	J	J	D
J	H	J	D	D	H

D has 22% + 23% = 45% of the first-place votes. H has 15% + 29% + 7%= 51% of the first-place votes. J has 4% of the first-place votes. Since H has the most first-place votes, Elizabeth Holtzman (H) is the winner by plurality voting.

23. (a) The only way an election in this system can go from having a unique winner to having a different unique winner is if the winning alternative in the first election has exactly two first-place votes, and one of these two voters changed his or her ballot by moving some other alternative into first place (yielding a worse outcome for this voter).

(b) This is a tie in the second election.

(c) Answers will vary.

25. (a) Assume that the winner with the voting paradox ballots is A. Consider the following two elections:

Election 1

Rank	Number of voters (3)		
First	A	B	C
Second	B	C	A
Third	C	A	B

Election 2

Rank	Number of voters (3)		
First	A	C	C
Second	B	B	A
Third	C	A	B

In Election 1, the winner is A (our assumption in this case) and in Election 2, the winner is C (because we are assuming that our voting system agrees with Condorcet's method when there is a Condorcet winner, as C is here). Notice that the voter in the middle, by a unilateral change in ballot, has improved the election outcome from his or her third choice to being his or her second choice. This is what that voter set out to do and is the desired instance of manipulation.

(b) Assume that the winner with the voting paradox ballots is B. Consider the following two elections:

Election 1

Rank	Number of voters (3)		
First	A	B	C
Second	B	C	A
Third	C	A	B

Election 2

Rank	Number of voters (3)		
First	A	B	A
Second	B	C	C
Third	C	A	B

In Election 1, the winner is B (our assumption in this case) and in Election 2, the winner is A (because we are assuming that our voting system agrees with Condorcet's method when there is a Condorcet winner, as A is here). Notice that the voter on the right, by a unilateral change in ballot, has improved the election outcome from his or her third choice to being his or her second choice. This is what that voter set out to do and is the desired instance of manipulation.

27. Properties 1 and 4.

29. Consider the following scenario: The chair votes for A and I vote for C. If you vote for B, the winner is A (your least preferred outcome) while the winner is C if you vote for C. This shows that voting for B does not weakly dominate your strategy of voting for C.

Chapter 11
Weighted Voting Systems

Exercise Solutions

1. **(a)** A winning or blocking coalition would be 50 senators plus the vice president, or more than 50 senators.

    (b) A three-fifths must include at least 60 senators. Thus it takes at least 41 senators to block ratification. The vice president cannot break a tie in this case, because a tie will not occur. Thus, the vice president is not a voter.

3. The weight-5 and weight-4 voters have veto power, because the coalition of all the voters has only 3 extra votes, less than they have. The weight-3 voter is a dummy, because the only winning coalition he or she he belongs to is the coalition with all the voters, and it has 3 extra votes.

5. **(a)** In 1958, B; G, and L had 5 votes between them. To pass a measure, 11 votes in addition to theirs would be needed. This would require participation of two of the remaining supervisors, who would not require any votes from B; G; or L to pass a measure. Thus, B, G, and L are dummies.

    In 1964 N, G and L have total weight of 25. To pass a measure an additional 33 is needed; this will take at least two of the other three voters. But any two of H_1, H_2, and B have total weight more than the quota, so they don't need the votes of N, G, or L. Thus, N, G, and L are dummies.

    In 1970 and 1976, G could team with H_1 and H_2 to pass a measure that the two Hempstead supervisors couldn't pass alone; thus G was not a dummy. The other supervisors, who had at least the weight of G, were thus also not dummies.

    In 1982, $\{30; 22; 6; 7\}$ is a winning coalition with weight exactly 65; thus none of its members is a dummy. The voters not in this coalition have more weight than G, so are also not dummies.

    (b) In 1958 and 1964, the two Hempstead supervisors had total weight more than the quota. Combined, they were dictators. The other voters were dummies.

    In 1970 and 1976, The Hempstead supervisors' total weight was less than the quota, but together with the least weight voter, G, they could form a winning coalition. Hence, there were no dummies in those years. In 1982, $\{H_1, H_2, L\}$ formed a winning coalition with weight exactly 65, thus L (and N and B, with more weight than L) was not a dummy. G would need to combine with voters with weight at least 59 to pass a measure; this would be impossible without including the Hempstead supervisors. When these were included with G, still one more supervisor would be needed - but the additional supervisor, with at least 7 votes, could, with the Hempstead supervisors, pass a measure without G. Thus, G was a dummy in 1982.

7. The last juror in the permutation is the pivotal voter.

9. None of these voting systems have dictators, nor does anyone have veto power. Therefore the pivotal position in each permutation is in position 2 or 3. Let's call the voters A, B, C, and D. This weighted voting system can be written as $[q : w_A, w_B, w_C, w_D] = [q : 30, 25, 24, 21]$.

(a) $[q : w_A, w_B, w_C, w_D] = [52 : 30, 25, 24, 21]$

A is pivot in four permutations where he or she is in position 2, and in all six permutations where she is in position 3: that's 10 in all.

Permutations		Permutations
B A C D		B C A D
B A D C		B D A C
C A B D		C B A D
C A D B		C D A B
		D B A C
		D C A B

B is pivot in two positions where he or she is in position 2 and four permutations where he or she is in position 3. Thus, Voter B is a pivot in 6 permutations.

Permutations		Permutations
A B C D		A D B C
A B D C		D A B C
		C D B A
		D C B A

C has the same power as B, and D is a pivot in the remaining two permutations.

Permutations
B C D A
C B D A

The Shapley-Shubik power index of this weighted voting system is therefore the following.

$$\left(\tfrac{10}{24}, \tfrac{6}{24}, \tfrac{6}{24}, \tfrac{2}{24}\right) = \left(\tfrac{5}{12}, \tfrac{1}{4}, \tfrac{1}{4}, \tfrac{1}{12}\right)$$

(b) $[q : w_A, w_B, w_C, w_D] = [55 : 30, 25, 24, 21]$

Now A is pivotal in only two permutations where he or she is in position 2. Voter A is still pivotal in all permutations when in position 3. Thus, Voter A is a pivot in 8 permutations.

Permutations		Permutations
B A C D		B C A D
B A D C		B D A C
		C B A D
		C D A B
		D B A C
		D C A B

B now has the same voting power as A. C and D are also equally powerful. Each is pivot in four permutations in which he or she is in third position and not preceded by A and B. Thus, the Shapley-Shubik power index of this weighted voting system is the following.

$$\left(\tfrac{8}{24}, \tfrac{8}{24}, \tfrac{4}{24}, \tfrac{4}{24}\right) = \left(\tfrac{1}{3}, \tfrac{1}{3}, \tfrac{1}{6}, \tfrac{1}{6}\right)$$

(c) $[q : w_A, w_B, w_C, w_D] = [58 : 30, 25, 24, 21]$

Any three voters have enough votes to win, and no two can win. The voters have equal power and the Shapley-Shubik power index of this weighted voting system is therefore $\left(\tfrac{1}{4}, \tfrac{1}{4}, \tfrac{1}{4}, \tfrac{1}{4}\right)$.

11. Bush's margin in Nevada would be $\dfrac{414,939-1000}{393,372+1000}=1.0496$.   This would move Nevada past Ohio in the permutations. Now Ohio's votes bring the total to 269 for Bush-Cheney, less than the quota. Nevada's 5 votes raises the total to 274, more than the quota. Therefore Nevada would be the pivot.

13. **(a)** We can represent a "yes" with 1, and a "no" with 0. Then the voting combinations are the 16 four-bit binary numbers: 0000, 0001, 0010, 0011, 0100, 0101, 0110, 0111, 1000, 1001, 1010, 1011, 1100, 1101, 1110, 1111. This would correspond to NNNN, NNNY, NNYN, NNYY, NYNN, NYNY, NYYN, NYYY, YNNN, YNNY, YNYN, YNYY, YYNN, YYNY, YYYN, YYYY.

(b) $\{\ \}$, $\{D\}$, $\{C\}$, $\{C,D\}$, $\{B\}$, $\{B,D\}$, $\{B,C\}$, $\{B,C,D\}$, $\{A\}$, $\{A,D\}$, $\{A,C\}$, $\{A,C,D\}$, $\{A,B\}$, $\{A,B,D\}$, $\{A,B,C\}$, and $\{A,B,C,D\}$.

(c) If the first bit of a given permutation is 1, then A votes "yes". If the second bit is 1, B votes "yes" in the corresponding coalition. The third bit tells us how C votes, and the fourth indicates the vote of D.

(d) i.   1
ii.   4
iii.   6

15. Let's call the participants A, B, C, and D in order of decreasing weight. This weighted voting system can be written as $\left[q:w_A,w_B,w_C,w_D\right]=\left[51:30,25,24,21\right]$.

The winning coalitions are those whose weights sum to 51 or more.

Winning coalition	Weight	Extra votes	Critical votes A	B	C	D
$\{A, B, C, D\}$	100	49	0	0	0	0
$\{A, B, C\}$	79	28	1	0	0	0
$\{A, B, D\}$	76	25	1	0	0	0
$\{A, C, D\}$	75	24	1	0	0	0
$\{B, C, D\}$	70	19	0	1	1	1
$\{A, B\}$	55	4	1	1	0	0
$\{A, C\}$	54	3	1	0	1	0
$\{A, D\}$	51	0	1	0	0	1
			6	2	2	2

Any voter that has a weight that exceeds the number of extra votes will be critical to that coalition. The critical voters in a coalition are indicated by a 1 in the table above.

A has a critical vote in 6 coalitions; B, C, and D each have critical votes in 2. Doubling to account for blocking coalitions, the Banzhaf power index is $(12,4,4,4)$.

17. (a) $_7C_3 = \dfrac{7!}{3!(7-3)!} = \dfrac{7!}{3!4!} = \dfrac{7\times 6\times 5}{3\times 2\times 1} = 7\times 5 = 35.$

(b) We can't use the formula that applied in part (a) because we'd get $\frac{50!}{100!(-50)!}$ and factorials of negative numbers are not defined. But really, the definition is all we need. If there are 50 voters, how many coalitions are there with 100 "yes" votes? NONE. The answer is $_{50}C_{100} = 0.$

(c) $_{15}C_2 = \dfrac{15!}{2!(15-2)!} = \dfrac{15!}{2!13!} = \dfrac{15\times 14}{2\times 1} = 15\times 7 = 105.$

(d) By the duality formula, $_{15}C_{13} = {}_{15}C_2 = 105.$ by the result of part (c).

19. In 1958 and 1964, Hempstead would dictate; all of the other supervisors would be dummies. The Hempstead supervisors (think as one voter with weight 18 or 62, depending on the year) are critical voters in $_4C_0 + {}_4C_1 + {}_4C_2 + {}_4C_3 + {}_4C_4$ coalitions. Performing this calculation (or using Pascal's triangle), we have the following.

$$_4C_0 + {}_4C_1 + {}_4C_2 + {}_4C_3 + {}_4C_4 = 1+4+6+4+1 = 16$$

Thus, the Hempstead supervisors are critical voter in 16 coalitions; doubling this, we find that their Banzhaf power index is 32.

Year	Banzhaf index
1958	$(32,0,0,0,0)$
1964	$(32,0,0,0,0)$

Notice also that the dictator in an n-voter system is critical in each voting combination; everyone else is a dummy. Noting that $2^5 = 32$, we can readily see that the Banzhaf index is $(32,0,0,0,0)$.

In 1970 and 1976, the Hempstead supervisors need at least one other voter. Thus, there would still be no dummies. The Hempstead supervisors (think as one voter with weight 62 or 70, depending on the year) are critical voters in $_4C_1 + {}_4C_2 + {}_4C_3 + {}_4C_4$ coalitions. Performing this calculation (or using Pascal's triangle), we have the following.

$$_4C_1 + {}_4C_2 + {}_4C_3 + {}_4C_4 = 4+6+4+1 = 15$$

Thus, the Hempstead supervisors are critical voter in 15 coalitions; doubling this, we find that their Banzhaf power index is 30. This can also be determined through realizing that there were just two voting combinations in which the Hempstead supervisors were not critical: when it voted "yes" and the others all voted "no" and when there is a unanimous "no" vote.

Each of the remaining voters are only critical once. Thus, each of the other voters have a Banzhaf power index of 2.

Year	Banzhaf index
1970	$(30,2,2,2,2)$
1976	$(30,2,2,2,2)$

Continued on next page

19. continued

In 1982, the Hempstead supervisors need at least 1 of the three voters (N, B, or L) to have a motion to pass. G would be a dummy. Since G can be there or not, the Hempstead supervisors (think as one voter with weight 58) are critical voters in twice ${}_3C_1 + {}_3C_2 + {}_3C_3$ coalitions. Performing this calculation (or using Pascal's triangle), we have the following.

$$ {}_3C_1 + {}_3C_2 + {}_3C_3 = 3 + 3 + 1 = 7 $$

Thus, the Hempstead supervisors are critical voters in 14 coalitions; doubling this, we find that their Banzhaf power index is 28. If G joins any of the coalitions, it wouldn't make a difference. N, B, and L are each critical in $2 \times 1 = 2$ coalitions. Thus, N, B, and L each have a Banzhaf power index of 4.

Year	Banzhaf index
1982	$(28, 4, 4, 0, 4)$

21. A juror J will be critical in two situations: A 5-1 decision with J in the majority and a 4-2 blocked decision with J one of the blockers. There are ${}_5C_4 = 5$ ways that J could get 4 other jurors to join in a 5-1 decision, and there are ${}_5C_1 = 5$ other jurors who could join J to block. Thus, there is a total of 10 voting combinations where J is critical, out of a total of $2^6 = 64$ voting combinations in all. Because the jurors are tossing coins, the probability of each combination is $\frac{1}{64}$. The probability that J will cast a critical vote is $\frac{10}{64} = \frac{5}{32}$.

23. (a) $\{A, C, D\}$ and $\{A, B\}$

(b) A belongs to each winning coalition, so if A opposes a motion it will not pass. A has veto power. There are no other minimal blocking coalitions that include A, but we may notice that every winning coalition contains either B or C and D. Thus, if B can combine forces with either C or D to defeat a motion, $\{B, C\}$ and $\{B, D\}$ are also minimal blocking coalitions.

(c) A has veto power and thus is a critical voter in all 5 of the winning coalitions. B is critical in 3 winning coalitions: $\{A, B, C\}$, $\{A, B, D\}$, and $\{A, B\}$. Finally, C and D are only critical in one coalition: $\{A, C, D\}$. The Banzhaf power index is $(10, 6, 2, 2)$.

(d) $[q : w_A, w_B, w_C, w_D] = [6 : 4, 2, 1, 1]$ is one set of weights that works, but there are many other solutions such as $[q : w_A, w_B, w_C, w_D] = [5 : 3, 2, 1, 1]$. One can reason that A, the only voter with veto power, must have the most votes, while B is more powerful than C or D (who are equally powerful).

(e) A will pivot in any permutation in which he or she comes after B or after C and D. He or she automatically pivots in the 6 permutations where he or she is in position 4, and also the 6 permutations where he or she is in position 3 because if B is not last in such a permutation, then he or she comes before A, and if Voter A is last, then C and D come before A. There are two permutations, $BACD$ and $BADC$ where A pivots in position 2. This adds up to $2 + 6 + 6 = 14$ pivots for A. D pivots in permutations where A and C appear before him or her, and B is last. There are 2 such permutations: $ACDB$ and $CADB$. C has the same number of pivots as D. We have accounted for $14 + 2 + 2 = 18$ permutations. The remaining 6 belong to B. The Shapley-Shubik index is $\left(\dfrac{14}{24}, \dfrac{6}{24}, \dfrac{2}{24}, \dfrac{2}{24} \right) = \left(\dfrac{7}{12}, \dfrac{1}{4}, \dfrac{1}{12}, \dfrac{1}{12} \right)$.

25. Let's call the chairperson C, and the other members X_1, X_2, X_3, and X_4. The minimal winning and blocking coalitions have the form $\{C, X_i\}$, for $i = 1, 2, 3, 4$, and $\{X_1, X_2, X_3, X_4\}$ is a winning coalition because the chairperson can't block a motion unless another member joins him or her. Thus, C doesn't have veto power. With the proposed weights, each of these coalitions has weight 4, and it follows that all winning coalitions have weight at least 4. Losing coalitions (the chair alone, or 3 or fewer of the other members) would have weight less than 4. Thus the voting system is equivalent to the one with the proposed weights, $[4 : 3, 1, 1, 1, 1]$.

27. (a) Give the dean 2 votes, and each faculty member 1 vote. The quota would have to be 4. We would have $\left[q : w_D, w_{F_1}, w_{F_2}, w_{F_3} \right] = [4 : 2, 1, 1, 1]$.

(b) Give the dean and the provost each 2 votes (they have equal power on the committee), and each faculty member 1 vote. The quota is 6. We would have the following.

$$\left[q : w_D, w_P, w_{F_1}, w_{F_2}, w_{F_3} \right] = [6 : 2, 2, 1, 1, 1]$$

29. An administrator casts a critical vote in any coalition that includes one other administrator and three or four faculty members. There are 2 ways to choose the other administrator, and 5 ways to choose the group of faculty members: 10 winning coalitions in which the administrator is critical.

A faculty member is critical in any coalition that includes exactly 2 other faculty members, and 2 or 3 administrators. There are 3 ways to choose the other faculty members, and 4 ways to assemble 2 or 3 administrators: 12 winning coalitions in which the faculty member is critical. The Banzhaf power index is $(24, 24, 24, 24, 20, 20, 20)$. The administrators probably think they are more powerful, but actually they aren't.

31. A voter has veto power if and only if he or she belongs to every winning coalition. Because each winning coalition contains at least one minimal winning coalition (you can obtain a minimal winning coalition by removing non-critical voters, one at a time, until there are no more), a voter who belongs to all minimal winning coalitions has veto power. If there is only one minimal winning coalition, then every voter in that coalition has veto power, and every voter who does not belong is a dummy. If there are only two minimal winning coalitions, since they overlap, at least one voter belongs to both and thus has veto power.

33. (a) Let's call the voters A, B, C, and D. We will omit E from the list. One could include him in any winning or losing coalition without altering the vote total. This weighted voting system can be written as $\left[q : w_A, w_B, w_C, w_D \right] = [51 : 48, 23, 22, 7]$.

Winning coalition	Weight	Extra votes	Losing coalition	Weight	Votes needed
$\{A, B, C, D\}$	100	49	$\{A\}$	48	3
$\{A, B, C\}$	93	42	$\{B, C\}$	45	6
$\{A, B, D\}$	78	27	$\{B, D\}$	30	21
$\{A, C, D\}$	77	26	$\{C, D\}$	29	22
$\{A, B\}$	71	20	$\{B\}$	23	28
$\{A, C\}$	70	19	$\{C\}$	22	29
$\{A, D\}$	55	4	$\{D\}$	7	44
$\{B, C, D\}$	52	1	$\{\ \}$	0	51

(b) A cannot sell more than 4 shares to B, because that is all the extra votes of $\{A, D\}$. The right column of the table in part (a) indicates that all of the losing coalitions involving B need more than 4 votes, so no additional winning coalitions would be created.

Continued on next page

33. continued

    (c) When selling to D, the extra votes of $\{A, D\}$ are unaffected. A can sell 19 shares to D. without changing any winning coalitions. The strongest losing coalition involving D needs 21 votes, and its status would not be affected by the sale.

    (d) A is again limited to the extra votes of $\{A, D\}$, 4 shares. Before starting, 8 new winning coalitions should be created, by combining E and each of the winning coalitions, and 8 losing coalitions would be created, combining E with the previous losing coalitions. No vote counts would change, because E has 0 shares at the outset. The sale would not affect the losing coalition E; it would still need 3 shares to win. The first losing coalition that would gain shares is $\{B, C, E\}$, which would still need 2 more shares to win.

35. **(a)** The ordinary members are equally powerful, so each gets 1 vote. The quota is 8, to make the coalition of all ordinary members winning, but 7 members losing. The chair gets 6 votes, enough to combine with 2 ordinary members and win. In our notation, the weighted voting system is $\left[q : w_C, w_{O_1}, w_{O_2}, w_{O_3}, w_{O_4}, w_{O_5}, w_{O_6}, w_{O_7}, w_{O_8} \right] = \left[8 : 6, 1, 1, 1, 1, 1, 1, 1, 1 \right]$.

    (b) The chairperson is critical in all winning coalition she belongs to, except the one in which the committee is unanimous. The number of these coalitions is $2^8 - {_8}C_0 - {_8}C_1 - {_8}C_8 = 256 - 1 - 8 - 1 = 246$, because there are 2^8 coalitions of ordinary members in all, of which we must eliminate ${_8}C_0 + {_8}C_1$ because they consist of 0 or 1 members, who cannot form a winning coalition with the chairperson, and ${_8}C_8$, because when all 8 ordinary members join the chairperson, the chairperson isn't critical. An ordinary member is critical in 8 winning coalitions: when joined by the rest of the ordinary members, and when joined by the chairperson and one of the other 7 ordinary members. Counting an equal number of blocking coalitions, the Banzhaf power index of this system is $\left(492, 16, 16, 16, 16, 16, 16, 16, 16 \right)$.

    (c) Divide the permutations into 9 groups, according to the location of the chairperson. She is pivot in groups 3, 4, 5, 6, 7, and 8. Therefore his or her Shapley-Shubik power index is $\frac{6}{9} = \frac{2}{3}$. Each ordinary member has $\frac{1}{8}$ of the remaining $1 - \frac{2}{3} = \frac{1}{3}$ of the power; hence the Shapley-Shubik power index of this system is as follows.

$$\left(\frac{2}{3}, \frac{1}{24}, \frac{1}{24}, \frac{1}{24}, \frac{1}{24}, \frac{1}{24}, \frac{1}{24}, \frac{1}{24}, \frac{1}{24} \right)$$

    (d) In this system, the chairperson is 30.75 times as powerful as an ordinary member according to the Banzhaf index, but only 16 times as powerful by the Shapley-Shubik power index.

37. Let's determine the minimal winning coalitions. They would be of the following types:

(a) 3 city officials

(b) 2 city officials and 1 borough president

(c) 1 city official and all of the borough presidents.

Thus, the city officials all have the same power, and the borough presidents, although weaker than the city officials, also have equal power. We will assign a voting weight of 1 to each borough president. Let C denote the voting weight of a city official and let q be the quota. To make the coalition of type (i) win, and 2 city officials lose, we have

$$2C < q \leq 3C.$$

To make coalitions of type (ii) win, we require $2C + 1 \geq q$. Combining these inequalities, we see that (if C is an integer), $q = 2C + 1$. The 5 borough presidents plus one city official can win, but 4 borough presidents plus a city official is a losing coalition: therefore

$$C + 4 < q \leq C + 5$$

and hence $q = C + 5$. We now have two expressions for q, $2C + 1$ and $C + 5$. Equating them, $2C + 1 = C + 5$, which we can solve for C to obtain $C = 4$, and hence $q = 9$. Finally, the 5 borough presidents form a losing coalition, but win if joined by a city official: this will hold provided

$$5 < q \leq C + 5$$

This is also valid for $q = 9$ and $C = 4$.

The weighted voting system is $\left[q : w_M, w_C, w_{CCP}, w_{P_1}, w_{P_2}, w_{P_3}, w_{P_4}, w_{P_5} \right] = [9 : 4, 4, 4, 1, 1, 1, 1, 1]$.

39. The three weight-3 voters, or 2 weight-3 voters and one weight-1 voter form minimal winning coalitions.

A weight-3 voter, A, is critical in any winning coalition with 7, 8, or 9 votes. There are 6 weight-7 coalitions that include A, because they are formed by assembling one of the other 2 weight-3 voters, and one of the 3 weight-1 voters. There are also 6 weight-8 coalitions with A: they also need one of the other 2 weight-3 voters and 2 of the 3 weight-1 voters (the number of ways to choose 2 weight-1 voters is $_3C_2 = 4$). Finally, there are 3 coalitions of weight 9 to which A belongs: all 3 weight-3 voters is one of them; the other 2 consist of A and one of the other 2 weight-3 voters, and all of the weight-1 voters. Thus A is critical in a total of 15 winning coalitions, and A's Banzhaf power index is 30.

A weight-1 voter, D, is critical in 3 winning coalitions, formed by assembling D with 2 of the 3 weight-3 voters. Doubling, we see that the Banzhaf power index of D is 6.

The Banzhaf power index of this system is $(30, 30, 30, 6, 6, 6)$.

41. Let's start with a weight-1 voter, A.

Case I: 3 weight-1 voters

A will be pivotal in permutations where he or she is in position 3, and positions 1 and 2 are occupied by weight-3 voters. There are $_3C_2 = 3$ ways to choose the weight-3 voters who come first, 2 ways to put them in order, and 3! ways to put the voters following A in order. Thus the Shapley-Shubik power index of A is $\dfrac{3 \times 2 \times 3!}{6 \times 5 \times 4 \times 3!} = \dfrac{3 \times 2}{6 \times 5 \times 4} = \dfrac{1}{5 \times 4} = \dfrac{1}{20}$.

The other weight-1 voters have the same power, and the weight-3 voters share the remaining $1 - 3 \times \dfrac{1}{20} = 1 - \dfrac{3}{20} = \dfrac{17}{20}$ of the power. Thus, each weight-3 voter has $\dfrac{1}{3}\left(\dfrac{17}{20}\right) = \dfrac{17}{60}$ of the power. The Shapley-Shubik index becomes $\left(\dfrac{17}{60}, \dfrac{17}{60}, \dfrac{17}{60}, \dfrac{1}{20}, \dfrac{1}{20}, \dfrac{1}{20}\right)$ and the weight-1 voter is $\dfrac{17}{60} \div \dfrac{1}{20} = \dfrac{17}{60} \cdot \dfrac{20}{1} = \dfrac{17}{3} = 5\dfrac{2}{3}$ times as much power.

Case II: 4 weight-1 voters

Now A will be pivotal in permutations where he or she is in position 3, and positions 1 and 2 are occupied by weight-3 voters. There are still 3 ways to select the 2 weight-3 voters and 2 ways to put them in order, but now there are 4! ways to arrange the voters who follow A in the permutation. This gives $6 \times 4!$ permutations in which A is pivotal.

A will also be pivotal in any permutation where he or she is in position 5 and the final two positions are occupied by weight-3 voters. There are the same number of these permutations. The Shapley-Shubik power index for A is therefore $\dfrac{2 \times 6 \times 4!}{7 \times 6 \times 5 \times 4!} = \dfrac{2 \times 6}{7 \times 6 \times 5} = \dfrac{2}{7 \times 5} = \dfrac{2}{35}$. The other 3 weight-1 voters have the same power, and the remaining $1 - 4 \times \dfrac{2}{35} = 1 - \dfrac{8}{35} = \dfrac{27}{35}$ of the power belongs to the weight-3 voters. Each weight-3 voter has $\dfrac{1}{3}\left(\dfrac{27}{35}\right) = \dfrac{9}{35}$ of the power. The Shapley-Shubik index becomes $\left(\dfrac{9}{35}, \dfrac{9}{35}, \dfrac{9}{35}, \dfrac{2}{35}, \dfrac{2}{35}, \dfrac{2}{35}, \dfrac{2}{35}\right)$ and the weight-1 voter is $\dfrac{9}{35} \div \dfrac{2}{35} = \dfrac{9}{35} \cdot \dfrac{35}{2} = \dfrac{9}{2} = 4\dfrac{1}{2}$ times as much power.

Although each voter's share of power decreased proportionally in the Banzhaf model when a new voter joined the system, in this particular situation, each weight-1 voter's power, measured by the Shapley-Shubik model, increased when the new voter was included, because $\dfrac{2}{35} > \dfrac{1}{20}$.

Chapter 12
Electing the President

Exercise Solutions

1. Assume a distribution is skewed to the left. The heavier concentration of voters on the right means that fewer voters are farther from the median. Because there are fewer voters "pulling" the mean rightward, it will be to the left of the median. Likewise, a distribution skewed to the right will have a mean to the right of the median.

3. While there is no median position such that half the voters lie to the left and half to the right, there is still a position where the middle voter (if the number of voters is odd) or the two middle voters (if the number of voters is even) are located, starting either from the left or right. In the absence of a median, less than half the voters lie to the left and less than half to the right of this middle voter's (voters') position (positions).

   Hence, any departure by a candidate from a position of a middle voter to the position of a non-middle voter on the left or right will result in that candidate's getting less than half the votes – and the opponent's getting more than half. Thus, the middle position (positions) is (are) in equilibrium, making it (them) the extended median.

5. When the four voters on the left refuse to vote for a candidate at 0.6, his opponent can do better by moving to 0.7, which is worse for the dropouts.

7. The voters are spread from 0.1 to 0.9, so it is a position at 0.5 that minimizes the maximum distance (0.4) a candidate is from a voter. If the candidates are at the median of 0.6, the voter at 0.1 would be a distance of 0.5 from them. In this sense, the median is worse than the mean of 0.56, which would bring the candidates closer to the farthest-away voter and, arguably, be a better reflection of the views of the electorate.

9. The middle peak will be in equilibrium when it is the median or the extended median. Yes, it is possible that, say, the peak on the left is in equilibrium, as illustrated by the following discrete-distribution example, in which the median is 0.2:

Position i	1	2	3	4	5	6	7
Location (l_i) of position i	0.1	0.2	0.3	0.5	0.6	0.8	0.9
Number of voters (n_i) at position i	7	8	1	2	1	2	1

11. If the population is not uniformly distributed and, say, 80% live between $\frac{3}{8}$ and $\frac{5}{8}$ and only 10% live to the left of $\frac{3}{8}$ and 10% to right of $\frac{5}{8}$, then the bulk of the population will be well served by two stores at $\frac{1}{2}$. In fact, stores at $\frac{1}{4}$ and $\frac{3}{4}$ will be farther away for 80% of the population, so it can be argued that the two stores at $\frac{1}{2}$ provide a social optimum.

13. Presumably, the cost of travel would have to be weighed against how much lower more competitive prices are.

15. Since the districts are of equal size, the mayor's median or extended median must be between the leftmost and rightmost medians or extended medians; otherwise, at least $\frac{2}{3}$ of the voters would be on one side of the mayor's position, which would preclude it from being the median or extended median. This is not true of the mean, however, if, say, the left-district positions are much farther away from the mayor's median or extended median than the right-district positions. In such a case, the mayor's mean would be in the interval of the left-district positions.

17. If, say, A takes a position at M and B takes a position to the right of M, C should take a position just to the left of M that is closer to M than B's position, giving C essentially half the votes and enabling him or her to win the election. If neither A nor B takes a position at M, C should take a position next to the player closer to M; the position that C takes to maximize his or her vote may be either closer to M (if the candidates are far apart) or farther from M (if the players are closer together), but this position may not be winning. For example, assume the voters are uniformly distributed over [0,1]. If $\frac{3}{16}$ of the voters lie between A (to the left of M) and M, and $\frac{3}{16}$ of the voters lie between M and B (to the right of M), then C does best taking a position just to the left of A or just to the right of B, obtaining essentially $\frac{5}{16}$ of the vote. To be specific, assume C moves just to the left of A. Then A will obtain $\frac{3}{16}$ of the vote, but B will win with $\frac{1}{2}$ of the vote (that to the right of M), so C's maximizing position will not always be sufficient to win.

19. Following the hint, C will obtain $\frac{1}{3}$ of the vote by taking a position at M, as will A and B, so there will be a three-way tie among the candidates. Because a non-unimodal distribution can be bimodal, with the two modes close to M, C can win if he or she picks up most of the vote near the two modes, enabling C to win with more than $\frac{1}{3}$ of the vote.

21. B should enter just to the right of $\frac{3}{4}$, making it advantageous for C to enter just to the left of A, giving C essentially $\frac{1}{4}$ of the vote. With C and A almost splitting the vote to the left of M and a little beyond, B would win almost all the vote to the right of M. (If C entered at $\frac{1}{2}$, he or she would get slightly more than $\frac{1}{4}$ of the vote but lose to A, who would get $\frac{3}{8}$.)

23. If the distribution is uniform, these positions are $\frac{1}{6}$, $\frac{5}{6}$, and $\frac{1}{2}$ for A, B, and C, respectively, making D indifferent between entering just to the left of A, just to the right of B, or in between A and C at $\frac{1}{3}$ or between C and B at $\frac{2}{3}$, which would give D $\frac{1}{6}$ of the vote in any case.

25. No, because Gore would get 49%, the same as Bush, so instead of winning Gore would tie with Bush.

27. It seems far too complicated a "solution" for avoiding effects caused by the Electoral College. Why not just abolish the Electoral College?

29. By definition, more voters prefer the Condorcet winner to any other candidate. Thus, if the poll identifies the Condorcet winner as one of the top two candidates, he or she will receive more votes when voters respond to the poll by voting for one or the other of these candidates. The possibility that the Condorcet winner might not be first in the poll, but win after the poll is announced, shows that the plurality winner may not be the Condorcet winner. Some argue that the Condorcet winner is always the "proper" winner, but others counter that a non-Condorcet winner who is, say, everybody's second-most-preferred candidate is a better social choice than a 51%-Condorcet winner who is ranked last by the other 49%.

31. D is the Condorcet winner. It is strange in the sense that a poll that identifies either the top two or the top three candidates would not include D.

33. A would win with 4 votes to 3 votes for B and 3 votes for C. It is strange that the number of top contenders identified by a poll can result in opposite outcomes (A in this exercise, whereas B defeats A when only two top contenders are identified by a poll, as in Exercise 32).

35. Assume a voter votes for just a second choice. It is evident that voting for a first choice, too, can never result in a worse outcome and may sometimes result in a better outcome (if the voter's vote for a first choice causes that candidate to be elected).

37. Following the hint, the voter's vote for a first and third choice would elect either A or C. If the voter also voted for B, then it is possible that if A and B are tied for first place, then B might be elected when the tie is broken, whereas voting for just A and C in this situation would elect A.

39. No. Voting for a first choice can never hurt this candidate and may help elect him or her.

41. No. If class I and II voters vote for all candidates in their preferred subsets, they create a three-way tie among A, B, and C. To break this tie, it would be rational for the class III voter to vote for both D and C and so elect C, whom this voter prefers to both A and B. But now class I voters will be unhappy, because C is a worst choice. However, these voters cannot bring about a preferred outcome by voting for candidates different from A and B.

43. Without polling, A in case (i), D in case (ii), and B and D in case (iii); with polling, B in case (i), D in case (ii), and D in case (iii).

45. Exactly half the votes, or 9.5 votes each.

47. Substitute into the formula for r_i, in Exercise 46, $d_i = (n_i / N) D$ and $D = R$. The proportional rule is "strategy-proof" in the sense that if one player follows it, the other player can do no better than to follow it. Hence, knowing that an opponent is following the proportional rule does not help a player optimize against it by doing anything except also following it.

49. To win in states with more than half the votes, any two states will do. Thus, there is no state to which a candidate should not consider allocating resources. In the absence of information about what one's opponent is doing, all states that receive allocations should receive equal allocations since all states are equally valuable for winning.

51. The Democrat can win the election by winning in any two states or in all three. The first three expressions in the formula for PWE_D give the probabilities of winning in the three possible pairs of states, whereas the final expression gives the probability of winning in all three states.

53. Yes, but in a complicated way. Intuitively, the large states that are more pivotal, and whose citizens therefore have more voting power (as shown in Chapter 11), are more deserving of greater resources (as shown in this chapter).

Chapter 13
Fair Division

Exercise Solutions

1. Donald initially receives the Palm Beach mansion (40 points) and the Trump Tower triplex (38 points) for a total of 78 points. Ivana initially receives the Connecticut estate (38 points) and the Trump Plaza apartment (30 points) for a total of 68 points. Because Ivana has fewer points than Donald, she receives the cash and jewelry (on which they both placed 2 points) bringing her total to 70 points. As Donald still has more points (78 to 70), we begin transferring items from him to her. To determine the order of transfer, we must calculate the point ratios of the items that Donald now has.

    The point ratio of the Palm Beach mansion is $\frac{40}{20} = 2.0$.

    The point ratio of the Trump Tower triplex is $\frac{38}{10} = 3.8$.

    Because $2.0 < 3.8$, the first item to be transferred is the Palm Beach mansion. However, if all of it were given to Ivana, her point total would rise to $70 + 20 = 90$, and Donald's point total would fall to $78 - 40 = 38$. This means that only a fraction of the Palm Beach mansion will be transferred from Donald to Ivana.

    Let x be the fraction of the Palm Beach mansion that Donald retains, and let $1-x$ be the fraction of it that is given to Ivana. To equalize point totals, x must satisfy $38 + 40x = 70 + 20(1-x)$.

    Thus, using algebra to solve this equation yields the following.

    $$38 + 40x = 70 + 20 - 20x$$
    $$38 + 40x = 90 - 20x$$
    $$60x = 52$$
    $$x = \tfrac{52}{60}$$
    $$x = \tfrac{13}{15}$$

    Thus Donald receives the Trump Tower triplex and $\frac{13}{15}$ (about 87%) ownership of the Palm Beach mansion for a total of about 72.7 of his points, and Ivana gets the rest (for about 72.7 of her points).

3. Mike initially gets his way on the room party policy (50), the cleanliness issue (6), and lights-out time (10) for a total of 66 points. Phil initially gets his way on the stereo level issue (22), smoking rights (20), phone time (8), and the visitor policy (5) for a total of 55 points. Because Phil has fewer points that Mike, he gets his way on the alcohol use issue, on which they both placed 15 points, bringing his total to 70. To determine the order of transfer (from Phil to Mike), we must calculate the point ratios of the issues on which Phil got his way.

    Point ratio of the stereo level issue is $\frac{22}{4} = 5.5$.

    Point ratio of the smoking rights issue is $\frac{20}{10} = 2.0$.

    Point ratio of the alcohol issue is $\frac{15}{15} = 1.0$.

    Point ratio of the phone time issue is $\frac{8}{1} = 8.0$.

    Point ratio of the visitor policy issue is $\frac{5}{4} = 1.25$.

    Continued on next page

3. (continued)

The first issue to be transferred is the alcohol issue, because it has the lowest point ratio. However, if all of it were given to Mike, his point total would rise to 66 + 15 = 81, and Phil's point total would fall to 70 – 15 = 55. This means that only a fraction of the alcohol issue will be transferred from Phil to Mike.

Let x be the fraction of the alcohol issues that Phil retains, and let $1-x$ be the fraction of it that is given to Mike. To equalize point totals, x must satisfy $55+15x = 66+15(1-x)$.

Thus, using algebra to solve this equation yields the following.

$$55+15x = 66+15-15x$$
$$55+15x = 81-15x$$
$$30x = 26$$
$$x = \tfrac{26}{30}$$
$$x = \tfrac{13}{15}$$

Thus, Phil gets his way on the stereo level issue, the smoking rights issue, the phone time issue, the visitor policy issue, and $\tfrac{13}{15}$ (about 87%) of his way on the alcohol issue for a total of 68 points. Mike gets his way on the rest.

5. Answers will vary.

7. Allocation 1:

    (a) Not proportional: Bob gets 10% in his eyes.

    (b) Not envy-free: Bob, for example, envies Carol.

    (c) Not equitable: Bob thinks he got 10% and Carol thinks she got 40%.

    (d) Example: Give Bob X, Carol Y, and Ted Z.

Allocation 2:

    (a) Not proportional: Carol gets 30% in her eyes.

    (b) Not envy-free: Carol, for example, envies Bob.

    (c) Not equitable: Bob thinks he got 50% and Carol thinks she got 30%.

    (d) Example: Give Bob Y, Carol X, and Ted Z.

Allocation 3:

    (a) Not proportional: Carol and Ted get 0% in their eyes.

    (b) Not envy-free: Carol and Ted envy Bob.

    (c) Not equitable: Bob thinks he got 100% and Carol thinks she got 0%.

    (d) It is Pareto optimal – for Carol or Ted to get anything, Bob will have to get less.

Allocation 4:

    (a) Not proportional: Carol gets 30% in her eyes.

    (b) Not envy-free: Carol, for example, envies Bob.

    (c) Not equitable: Bob thinks he got 50% and Carol thinks she got 30%.

Allocation 5:

    (a) It is proportional.

    (b) Not envy-free: Bob, for example, envies Carol.

    (c) It is equitable.

9. They handle the car first, as in Exercise #8. Then Mary gets the house and places $\frac{59,100}{2} = 29,550$ in a kitty. John takes out $\frac{55,900}{2} = 27,950$ and they split the remaining $29,550 - 27,950 = 1,600$ equally. Thus, for the house, Mary gets it and gives John $28,750. In total, Mary gets both the car and the house and pays John $15,081.25 + \$28,750 = \$43,831.25$.

11. First, C gets the house and places two-thirds of 165,000 (i.e., 110,000) in a kitty. A then withdraws one-third of 145,000 (i.e., 48,333) and B withdraws one-third of 149,999 (i.e., 50,000). They divide the remaining 11,667 equally among the three of them.

    Second, A gets the farm and places two-thirds of 135,000 (i.e., 90,000) in a kitty. B then withdraws one-third of 130,001 (i.e., 43,334) and C withdraws one-third of 128,000 (i.e., 42,667). They divide the remaining 3,999 equally among the three of them.

    Third, C gets the sculpture and deposits two-thirds of 127,000 (i.e., 84,667) in a kitty. A then withdraws one-third of 110,000 (i.e., 36,667) and B withdraws one-third of 80,000 (i.e., 26,667). They divide the remaining 21,333 equally among them.

    Thus, A gets the farm and receives $52,222 + \$43,778$ and pays $44,667 + \$44,000$, so A, in total, receives the farm plus $7,333. Similarly, B receives $132,334 and C receives both the house and the sculpture, while paying $139,667.

13. The table is as follows.

Potential Recipient	Months waiting, position in terms of waiting, points awarded	Antigens matched and points awarded	Percent sensitized and points awarded	Total points
A	9 (1) $10 \times \left(\frac{4}{4}\right) = 10$ points	2 $2 \times 2 = 4$ points	20 2 points	16 points
B	6 (2) $10 \times \left(\frac{3}{4}\right) = 7.5$ points	3 $2 \times 3 = 6$ points	0 0 points	13.5 points
C	5 (3) $10 \times \left(\frac{2}{4}\right) = 5$ points	4 $2 \times 4 = 8$ points	40 4 points	17 points
D	2 (4) $10 \times \left(\frac{1}{4}\right) = 2.5$ points	6 $2 \times 6 = 12$ points	60 6 points	20.5 points

15. One could use an absolute measure, awarding; for example, one point for each month a potential recipient has been waiting.

17. The bottom-up strategy fills in the blanks as follows:

    Carol: investments   boat   washer-dryer

    Bob: car   television   CD player

    Thus, Carol first chooses the investments, and the final allocation has her also receiving the boat and the washer-dryer.

19. The bottom-up strategy fills in the blanks as follows:

Fred: <u>boat</u>   <u>car</u>   <u>motorcycle</u>

Mark: <u>tractor</u>   <u>truck</u>   <u>tools</u>

Thus, Fred first chooses the boat, and the final allocation has him also receiving the car and the motorcycle.

21. The bottom-up strategy fills in the blanks as follows (CT stands for Connecticut):

Ivana: <u>CT estate</u>   <u>apartment</u>   <u>cash and jewelry</u>

Donald: <u>mansion</u>   <u>triplex</u>

Thus, Ivana first chooses the Connecticut estate, and the final allocation has her also receiving the Trump Plaza apartment and the cash and jewelry.

23. One way is to have Bob divide the cake into four pieces and to let Carol choose any three. Another is to have Bob divide the cake into two pieces and then let Carol choose one. Then they can do divide-and-choose on the piece that Carol did not choose.

25. (a) Bob should be the divider. That way, he can get 12 units of value instead of 9 units of value.

(b) Here, Bob knows the preferences of the other party. In Exercise 22, we assumed that the divider didn't know the preferences of the other party.

27. (a) See figures below.

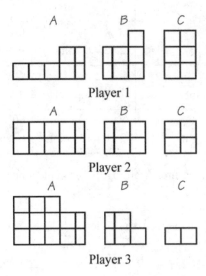

Player 1

Player 2

Player 3

Continued on next page

27. (continued)

    (b) Player 2 finds A acceptable (9 square units), but not B (5 square units) or C (4 square units). Player 3 finds A acceptable (12 square units), but not B (4 square units) or C (2 square units).

    (c) Players 2 and 3 both find B and C unacceptable. (C is on the right.)

    (d) (i) Assume C is given to Player 1. If Player 2 cuts the rest, he will make each piece 7 square units. Player 3 will choose the leftmost piece, which he thinks is 10 square units. Thus, Player 1 gets a piece he thinks is 6 square units. Player 2 gets a piece he thinks is 7 square units, and Player 3 gets a piece he thinks is 10 square units.

        (ii) If Player 3 cuts the rest, she will make each piece 8 square units. (This requires a vertical cut two-thirds of the way across the third triple of squares.) Player 2 will choose the rightmost piece, which she thinks is $8\frac{2}{3}$ square units. Thus, Player 1 gets a piece she thinks is 6 square units, Player 2 gets a piece she thinks is $8\frac{2}{3}$ square units, and Player 3 gets a piece she thinks is 8 square units.

29. (a) Ted thinks he is getting at least one-third of the piece that Bob initially received and at least one-third of the piece that Carol initially received. Thus, Ted thinks he is getting at least one-third of part of the cake (Bob's piece) plus one-third of the rest of the cake (Carol's piece).

    (b) Bob gets to keep exactly two-thirds (in his own view) of the piece that he initially received and thought was at least of size one-half. Two-thirds times one-half equals one-third.

    (c) If, for example, Ted thinks the "half" Carol initially gets is worthless, then Ted may wind up thinking that he (Ted) has only slightly more than one-third of the cake, while Bob has (in Ted's view) almost two-thirds of the cake. In such a case, Ted will envy Bob.

31. (a) If a player follows the suggested strategy, then clearly he or she will receive a piece of size exactly one-fourth *if* he or she does, in fact, call cut at some point. How could a player (Bob, for example) fail to call cut when using this strategy? Only if each of the other three players "preempted" Bob by calling cut before he did each time the knife was set in motion. But this means that each of the other three is left with a piece that Bob considered to be of size less than one-fourth. Hence, when the other three players have left with their shares, there is, in Bob's view, over one-fourth of the cake left for him.

    (b) If you call cut first – and thus exit the game with a piece of size exactly one-fourth in your estimation – you will envy the next player to receive a piece *if* no one calls cut until the next piece is larger than one-fourth in your estimation.

    (c) If there are four players and the first player has exited with his or her piece, then you could wait to call cut until the knife reaches the point where one-half of the original cake is left. Alternatively, you could wait until the knife passed over one-third of what was left.

33. (a) The knife on the left would be at the point where the other knife started. (Thus, the portion between the knives would be the complement of the piece A.)

    (b) If, for example, Carol thinks the portion between the knives at the beginning (i.e., piece A) is of size less than one-half, then she definitely will think the portion between the two knives at the end (i.e., the complement of piece A) is of size greater than one-half. Because this portion of cake between the two knives goes from being of size less than one-half in her estimation to being of size greater than one-half in her estimation, there must be a point where it is of size exactly one-half in her estimation. An analogous argument applies if Carol thinks that A is of size greater than one-half.

Chapter 14
Apportionment

Exercise Solutions

1. Jane's total expenses are \$71. The calculation of the percentages is shown in the table.

	Percentage	rounded
Rent	$\frac{31}{71} \times 100\% = 43.66\%$	44%
Food	$\frac{16}{71} \times 100\% = 22.54\%$	23%
Transportation	$\frac{7}{71} \times 100\% = 9.86\%$	10%
Gym	$\frac{12}{71} \times 100\% = 16.90\%$	17%
Miscellaneous	$\frac{5}{71} \times 100\% = 7.04\%$	7%

The percentages add up to 101%.

3. The new enrollments are obtained by subtracting from the enrollment of each level the number of students who are moving to a lower level, and add to each the number of students who are moving from a higher level. Here are the calculations.

Calculus I	$500 + 45 =$	545
Calculus II	$100 - 45 + 41 =$	96
Calculus III	$350 - 41 + 12 =$	321
Calculus IV	$175 - 12 =$	163

The total number of students enrolled remains 1125, and the average number of students per teaching assistant is still 56.25. Here are the new quotas.

Calculus I	$545 \div 56.25 =$	9.69
Calculus II	$96 \div 56.25 =$	1.71
Calculus III	$321 \div 56.25 =$	5.71
Calculus IV	$163 \div 56.25 =$	2.90

The new rounded quotas are as follows.

Calculus I	10
Calculus II	2
Calculus III	6
Calculus IV	3
Total	21

This calls for too many teaching assistants, so the numbers must be adjusted. The apportionment methods introduced in this chapter present a variety of approaches to solving this problem but

5. Rounding each of the summands down, the sum of the lower quotas is $0 + 1 + 0 + 2 + 2 + 2 = 7$. The three numbers with the greatest fractional parts, 0.99, 1.59, and 2.38, receive their upper quotas. The apportioned sum is $0 + 2 + 1 + 2 + 3 + 2 = 10$.

7. The total number of votes cast is 510,000, so the standard divisor is $510,000 \div 102 = 5,000$. The quotas are obtained by dividing each state's population by this divisor, obtaining 50.8, 30.6, and 20.6, respectively. The lower quotas add up to 100, so we must increase the apportionment of two parties to their upper quotas. The Pro-UFO party, whose quota of 50.8 has the largest fractional part, gets an increase. The fractional parts of the quotas of the Anti-UFO and Who Cares parties are both equal to 0.6. They are tied for priority in receiving the last seat. A coin toss is probably the fairest way to settle this dispute.

9. The total enrollment is 115, and the standard divisor is 23. The quotas are as follows.

Geometry	$77 \div 23 =$	3.35 sections
Algebra	$18 \div 23 =$	0.78 sections
Calculus	$20 \div 23 =$	0.87 sections

The lower quota for geometry is 3, and the other two subjects have 0 lower quotas. Because they have larger fractional parts than geometry, they both receive their upper quotas, 1 each. The apportionment is as follows.

Geometry	3 sections
Algebra	1 section
Calculus	1 section

11. **(a)** The states in this apportionment problem are the investors, the seats are the 100 coins, and the populations are the individual investments. Thus, the standard divisor is $10,000 \div 100$ coins $= \$100$ per coin. The quotas, which represent the number of coins each investor should receive if fractional coins were possible, are obtained by dividing each investment by this divisor.

	Quota	Lower quota
Abe	36.190	36
Beth	18.620	18
Charles	22.580	22
David	20.100	20
Esther	2.510	2
Total	100.00	98

Two investors will receive their upper quotas: Beth and Charles, who have the largest fractions. Here are the apportionments, *before the excise tax was paid.*

Abe	36
Beth	19
Charles	23
David	20
Esther	2
Total	100

Continued on next page

11. continued

(b) When the excise tax is added, populations change, and the standard divisor changes as follows.

$$\$10,050 \div 100 = \$100.50 \text{ per coin}$$

We have to recalculate the quotas. The revised investments are divided by the new standard divisor as follows.

	Investment	Quota	Lower quota
Abe	$3,635	36.169	36
Beth	$1,864	18.547	18
Charles	$2,259	22.478	22
David	$2,042	20.318	20
Esther	$250	2.488	2
Totals	$10,050	100.000	98

Again, two investors will receive their upper quotas: Beth and Aunt Esther. The final apportionments are as follows.

	Before tax	After tax
Abe	36	36
Beth	19	19
Charles	23	22
David	20	20
Esther	2	3
Totals	100	100

(c) Aunt Esther not only got a dollar back, but Charles had to give her one of his rare coins! At least it's still in the family. The cause of this confusion is, of course, the population paradox.

13. Let's call the parties *A*, *B*, *C*, *D*, and *E*. The standard divisor for each house size, and the corresponding quotas are displayed in the following table. Each standard divisor is obtained by dividing the total number of votes cast in the election, 18,121,741, by the house size, either 89, 90, or 91. The quotas are determined by dividing the votes cast for the party by the standard divisor for the particular house size.

House size		89	90	91
Standard Divisor		203,615	201,353	199,140
Party	Votes		Quotas	
A	5,576,330	27.387	27.694	28.002
B	1,387,342	6.814	6.890	6.967
C	3,334,241	16.375	16.559	16.743
D	7,512,860	36.897	37.312	37.7276
E	310,968	1.527	1.544	1.562

Continued on next page

13. continued

The next table displays the lower quotas and their sum for each of the house sizes under consideration.

State	Lower Quotas		
A	27	27	28
B	6	6	6
C	16	16	16
D	36	37	37
E	1	1	1
Totals	86	87	88
Shortages	3	3	3

The last row of the above table records the number of seats that still must be apportioned. These seats go to the states whose quotas have the largest fractional parts. The final apportionments are as follows.

State	State Population	Apportionments		
A	5,576,330	27	28	28
B	1,387,342	7	7	7
C	3,334,241	16	17	17
D	7,512,860	37	37	38
E	310,968	2	1	1
Totals	18,121,741	89	90	91

The Alabama paradox occurs when the apportionment for the smallest state decreases from 2 to 1 as the house size increases from 89 to 90.

15. As with the Hamilton method, we have the following quotas.

Geometry	3.35 sections
Algebra	0.78 sections
Calculus	0.87 sections

The tentative apportionments are geometry, 3; algebra and calculus, 0. The critical divisors are determined by adding 1 to the tentative apportionments and dividing the result into the population of the subject, and are as follows.

Geometry	$77 \div 4 =$	19.25 students
Algebra	$18 \div 1 =$	18 students
Calculus	$20 \div 1 =$	20 students

Calculus has the greatest critical divisor, and its tentative apportionment is now 1. It receives a new critical divisor, $20 \div 2 = 10$. Now the greatest critical divisor is that of geometry, so its apportionment is 4. The house is full, and the Jefferson apportionment is

Geometry	4 sections
Algebra	cancelled!
Calculus	1 section

17. All three divisor methods start with the quotas, which were computed in Exercise 6.

	36 pearls	37 pearls
Abe	14.25	14.65
Beth	18.36	18.87
Charles	3.38	3.48

Jefferson method: The tentative apportionments are, for 36 or 37 pearls, Abe, 14; Beth, 18; and Charles, 3. With 36 pearls, 1 is left to be apportioned; with 37 there are 2 left. Here are the critical divisors.

Abe	$\$5,900 \div 15 =$	$\$393.33$
Beth	$\$7,600 \div 19 =$	$\$400.00$
Charles	$\$1,400 \div 4 =$	$\$350.00$

The 36th pearl goes to Beth. When the 37th pearl is discovered, there is no need to repeat the calculations. Beth's critical divisor (only) has to be recomputed, because she has another pearl now. Now her critical divisor is $\$7,600 \div 20 = \$380.00$. The highest priority for the 37th pearl goes to Abe. Here are the final Jefferson apportionments.

	36 pearls	37 pearls
Abe	14	15
Beth	19	19
Charles	3	3

Webster method: The tentative apportionments are obtained by rounding the quotas. With 36 pearls, all the quotas are rounded down, so the tentative apportionments add up to 35. We will have to calculate critical divisors to allocate the 36th pearl.

Abe	$\$5,900 \div 14.5 =$	$\$406.90$
Beth	$\$7,600 \div 18.5 =$	$\$410.81$
Charles	$\$1,400 \div 3.5 =$	$\$400.00$

Continued on next page

17. continued

Beth, with the greatest critical divisor, gets the 36th pearl. With 37 pearls, Abe's and Beth's quotas are both rounded up, and Charles's is rounded down. These tentative apportionments, 15, 19, and 3, add up to 37. Abe receives the 37th pearl. Here are the final Webster apportionments.

	36 pearls	37 pearls
Abe	14	15
Beth	19	19
Charles	3	3

The three methods produce the same results. If there is a principle on which to choose a method, it would probably be to choose the method by which the cost per pearl is as close as possible to the same for each of the friends. The cost per pearl is the district size. To minimize differences in district size, they should use the Dean method, which is the subject of Writing Project 4. This would be a good project for Charles, because it allocates the 36th pearl to Beth and the 37th to him!

19. The percentages are the quotas.

Hamilton method: Start with the lower quotas, $87 + 10 \times 1$, whose sum is 97. The three percentages with the greatest fractional parts, 87.85, 1.26, and 1.25, are rounded up to get the upper quotas; the remaining percentages are rounded down. The final apportionment is

$$88 + 2 + 2 + 1 + 1 + 1 + 1 + 1 + 1 + 1 + 1 = 100\%.$$

The first three percentages are rounded to upper quotas, and the remaining percentages are rounded to lower quotas. The quota condition is satisfied.

Jefferson method: Tentatively apportion to each percentage its lower quota. The critical divisors are then the unrounded percentage divided by (1 + the tentative apportionment). Thus, the critical divisor belonging to 87.85% is $87.85 \div 88 = 0.9983$, while the critical divisors belonging to the smaller percentages range from $1.26 \div 2 = 0.63$ down to $1.17 \div 2 = 0.585$. The largest critical divisor belongs to 87.85%, so its tentative apportionment is increased to 88 and its new critical divisor is $87.85 \div 89 = 0.9871$. This is still the largest critical divisor, so the apportionment of 87.85% is increased to 89. The new critical divisor, $87.85 \div 90 = 0.9761$, is still the largest, so its apportionment is increased to 90. Now the house is full, and the Jefferson apportionment is

$$90 + 1 + 1 + 1 + 1 + 1 + 1 + 1 + 1 + 1 + 1 = 100\%.$$

This apportionment rounds 87.85% to 90%, more than the upper quota. The quota condition is violated.

Webster method: The rounded percentages add up to 98, so we need to calculate critical divisors. The critical divisor belonging to 87.85% is $87.85 \div 88.5 = 0.9927$. Among the smaller percentages, the largest critical divisor is that of 1.26%, which is $1.26 \div 1.5 = 0.84$. The point goes to 87.85%, whose apportionment increases to 89. This calls for a new critical divisor, $87.85 \div 89.5 = 0.9816$, which exceeds the critical divisors of the smaller percentages. The apportionment of 87.85% is therefore increased again to 90. The final apportionment is the same as the Jefferson apportionment, so it too violates the quota condition.

21. Before the excise tax was included, the quotas, calculated as in Exercise 11, are rounded to obtain a tentative apportionment.

	Quota	Rounded quota
Abe	36.19	36
Beth	18.62	19
Charles	22.58	23
David	20.10	20
Esther	2.51	3
Total	100.00	101

One quota must be reduced, so we calculate critical divisors as follows.

Abe	$\$3619 \div (36 - 0.5) =$	$\$101.94$
Beth	$\$1862 \div (19 - 0.5) =$	$\$100.65$
Charles	$\$2258 \div (23 - 0.5) =$	$\$100.36$
David	$\$2010 \div (20 - 0.5) =$	$\$103.08$
Esther	$\$251 \div (3 - 0.5) =$	$\$100.40$

The least critical divisor is Charles's, so his apportionment is 22. After the tax is added, new rounded quotas are calculated.

	Quota	Rounded quota
Abe	36.17	36
Beth	18.55	19
Charles	22.48	22
David	20.32	20
Esther	2.49	2
Totals	100.01	99

Now one of the tentative apportionments must increase, so we must again compute critical divisors.

Abe	$\$3635 \div (36 + 0.5) =$	$\$99.589$
Beth	$\$1864 \div (19 + 0.5) =$	$\$95.590$
Charles	$\$2259 \div (22 + 0.5) =$	$\$100.400$
David	$\$2042 \div (20 + 0.5) =$	$\$99.610$
Esther	$\$250 \div (2 + 0.5) =$	$\$100.000$

Continued on next page

21. continued

Charles has the largest critical divisor, so his apportionment is increased to 23.   The final apportionments are as follows.

	Before tax	After tax
Abe	36	36
Beth	19	19
Charles	22	23
David	20	20
Esther	3	2
Totals	100	100

Esther must give one of her three rare coins to her nephew.

23.   Let's start by apportioning with 89 as the house size. The following table gets us started.

Party	Lower quota	Tentative apportionment
A	27.387	27
B	6.814	7
C	16.375	16
D	36.897	37
E	1.527	2
Totals	89	89

The quotas were calculated in the solution to Exercise 13, and the tentative apportionments were determined by rounding the quotas in the usual way, numbers with fractional parts being rounded to the lower quota if the fraction is less than 0.5, and to the upper quota otherwise.

Because the tentative apportionments add up to the house size, we are finished - they are the actual apportionments.

Moving to a 90-seat house, we will compare critical divisors, as shown in the following table.

Party	Votes	Tentative apportionment	Critical divisor
A	5,576,330	27	202,776
B	1,387,342	7	184,979
C	3,334,241	16	202,075
D	7,512,860	37	200,343
E	310,968	2	124,387
Totals	18,121,741	89	

Party A has first priority for another seat, so with a 90-seat house, its apportionment is 28.   We recompute its critical divisor as $5,576,630 \div 28.5 = 195,661$.   Thus party C now has the largest critical divisor, and gets the 91st seat, bringing its apportionment to 17.   The table below displays the apportionments (AP) in comparison with the rounded quotas (RQ) (obtained by rounding the quotas in the solution to Exercise 13).

House size	89		90		91	
Party	RQ	AP	RQ	AP	RQ	AP
A	27	27	28	28	28	28
B	7	7	7	7	7	7
C	16	16	17	16	17	17
D	37	37	37	37	38	37
E	2	2	2	2	2	2
Totals	89	89	91	90	92	91

There are no paradoxes or quota violations.

25. (a) One quota will be rounded up, and the other down to obtain the Webster apportionment. The quota that is rounded up will have fractional part greater than 0.5, and will be greater than the fractional part of the quota that is rounded down. The Hamilton method will give the party whose quota has the larger fractional part an additional seat. Thus the apportionments will be identical.

(b) These paradoxes never occur with the Webster method, which gives the same apportionment in this case.

(c) The Hamilton method, which always satisfies the quota condition, gives the same apportionment.

(d) Not always. Assume that parliament has 100 seats. If one party gets only 0.6% of the vote, and the other party gets 99.4%, the Jefferson critical divisor for the former party will be $\frac{0.6}{1}$, and the latter party will have a critical divisor of $\frac{99.4}{100}$. Jefferson would therefore apportion all 100 seats to the second party, since its critical divisor is the larger. Hamilton would apportion one seat to the first party.

27. Jim is 7 inches taller than Alice. The relative difference of their heights is 7 inches divided by Alice's height, 65 inches: $\frac{7}{65} = 10.77\%$.

29. (a) The absolute difference is $1.6114 - 1.1046 = 0.5068$. The relative difference is obtained by dividing the absolute difference by the smaller representative share, Montana's:

$$0.5068 \div 1.1046 = 45.88\%.$$

This is the same as the relative difference in district population.

(b) The representative shares are the reciprocals of the district populations, from Exercise 28, part (c). In units of seats per million, one obtains the representative share by dividing the average district population into one million (1,000,000). Thus the representative share for North Carolina would be $1,000,000 \div 672,306 = 1.4874$ seats per million, while Montana's representative share would $1,000,000 \div 452,658 = 2.2092$. The absolute difference, in Montana's favor, is 0.7218. The relative difference would be the same as that for the relative difference in district population for the same apportionment, $0.7218 \div 1.4874 = 48.53\%$.

(c) The apportionment giving North Carolina 13 seats, and Montana 1 seat yields a lesser absolute (and relative) difference in representative share than does the $12 - 2$ apportionment.

31. The rounding point between 0 and 1 is $\sqrt{0 \times 1} = 0$; the rounding point between 1 and 2 is $\sqrt{1 \times 2} = 1.4142$; the rounding point between 2 and 3 is $\sqrt{2 \times 3} = 2.4495$; and the rounding point between 3 and 4 is $\sqrt{3 \times 4} = 3.4641$.

33. The standard divisor is $(36+61+3) \div 5 = 20$ students. The quotas are as follows.

Algebra	$36 \div 20 =$	1.8 sections
Geometry	$61 \div 20 =$	3.05 sections
Calculus	$3 \div 20 =$	0.15 sections

Webster would round the quotas to 2, 3, and 0, respectively. These tentative apportionments add up to 5, the house size, and are the final Webster apportionments. Because Hill-Huntington rounds all numbers between 0 and 1 to 1, its tentative apportionment would be 2, 3, and 1. This would exceed the house size by 1, so we have to reduce one of the tentative apportionments. This requires critical divisors. They are as follows.

Algebra	$36 \div \sqrt{2 \times 1} =$	25.456 students
Geometry	$62 \div \sqrt{3 \times 2} =$	24.903 students
Calculus	$7 \div \sqrt{1 \times 0} =$	∞ students

The least critical divisor belongs to Geometry, so its apportionment is decreased to 2. In summary, here are the apportionments.

	Webster	Hill-Huntington
Algebra	2	2
Geometry	3	2
Calculus	cancelled!	1

It's likely that the principal would prefer the Webster method, because classes as small as the calculus class, with 3 students, should be cancelled. Notice that the Hill-Huntington apportionment gives Geometry less than its lower quota in order to accommodate Calculus.

35. Let's start by taking a seat from California, putting it in play. This leaves 52 seats for California, and California's priority for getting the extra seat is measured by its critical divisor,

$$\frac{\text{Population of California}}{\sqrt{52 \times 53}} = 646,330.227.$$

To secure the seat in play, Utah's population has to increase enough so that its critical divisor,

$$\frac{\text{Revised population of Utah}}{\sqrt{3 \times 4}},$$

surpasses California's. Thus, Utah needs a population of more than the following.

$$646,330.227 \times \sqrt{12} \approx 2,238,953.583$$

The 2000 census recorded Utah's population as 2,236,714, so an additional 2240 residents would be needed.

37. Let V_1, V_2, and V_3 be the district populations. The total population of the city is $V_1 + V_2 + V_3 = 1,400,000$; dividing by 20 we find the standard divisor is $s = 70,000$. The quotas for the districts are $q_1 = V_1 \div s = \frac{10}{7} \approx 1.429$, $q_2 = V_2 \div s = \frac{60}{7} \approx 8.571$, and $q_3 = V_3 \div s = 10$. We have $\lfloor q_1 \rfloor = 1$, so the first quota has rounding point $q_1^* = \sqrt{2} \approx 1.414$. Also, $\lfloor q_2 \rfloor = 8$, so the second rounding point is $q_1^* = \sqrt{8 \times 9} \approx 8.4853$. Note that $q_1 > q_1^*$ and $q_2 > q_2^*$ so the tentative Hill-Huntington apportionments of the first and second districts are $N_1 = \lceil q_1 \rceil = 2$ and $N_2 = \lceil q_2 \rceil = 9$. Of course, the tentative Hill-Huntington apportionment for the third district is $N_3 = 10$. Thus $2 + 9 + 10 = 21$ seats have been tentatively apportioned; we have to compare critical divisors. The critical divisor for the ith district is $d_i^- = V_i \div \sqrt{N_i(N_i - 1)}$. The critical divisors for the three districts are therefore as follows.

$$d_1^- = \frac{100,000}{\sqrt{2}} \approx 70,711$$

$$d_2^- = \frac{600,000}{\sqrt{72}} \approx 70,711$$

$$d_3^- = \frac{700,000}{\sqrt{90}} \approx 73,786$$

Here, the symbol $\approx$ means "approximately equal to."

The district with the smallest critical divisor must have its apportionment reduced to bring the total number of seats apportioned to 20. Our approximations do not tell us for sure how to compare d_1^- and d_2^-, but if we notice that $\sqrt{72} = \sqrt{6^2 \times 2} = 6\sqrt{2}$, then we will find that

$$d_2^- = \frac{600,000}{\sqrt{72}} = \frac{100,000}{\sqrt{2}} = d_1^-.$$

So, in fact, the first and second districts have equal critical divisors; and of course d_3^- is greater. Therefore there is a tie.

39. (a) Lowndes favors small states, because in computing the relative difference, the fractional part of the quota will be divided by the lower quota. If a large state had a quota of 20.9, the Lowndes relative difference works out to be 0.045. A state with a quota of 1.05 would have priority for the next seat.

(b) Yes, because like the Hamilton method, the Lowndes method presents a way to decide, for each state, if the lower or upper quota should be awarded.

(c) Yes. Since the method is not a divisor method, the population paradox is inevitable.

Continued on next page

39. continued

  (d) Let r_i denote the relative difference between the quota and lower quota for state i. The following table displays the numbers r_i for each state. Because the lower quotas add up to 97, the 8 states with the largest values in the r_i column will receive their upper quotas.

State	V_i	q_i	$\lfloor q_i \rfloor$	r_i	rank	a_i
Virginia	630,560	18.310	18	1.7%	14	18
Massachusetts	475,327	13.803	13	6.2%	8	14
Pennsylvania	432,879	12.570	12	4.8%	9	12
North Carolina	353,523	10.266	10	2.7%	13	10
New York	331,589	9.629	9	7.0%	7	10
Maryland	278,514	8.088	8	1.1%	15	8
Connecticut	236,841	6.877	6	14.6%	6	7
South Carolina	206,236	5.989	5	19.8%	5	6
New Jersey	179,570	5.214	5	4.3%	10	5
New Hampshire	141,822	4.118	4	3.0%	11	4
Vermont	85,533	2.484	2	24.2%	4	3
Georgia	70,835	2.057	2	2.9%	12	2
Kentucky	68,705	1.995	1	99.5%	1	2
Rhode Island	68,446	1.988	1	98.8%	2	2
Delaware	55,540	1.613	1	61.3%	3	2
Totals	3,615,920	105	97	–	–	105

41. (a) No. Unless a state's quota is a whole number, its tentative apportionment will be more than its quota. Of course, in the unlikely event that a state's quota is a whole number, its tentative apportionment would equal its quota. The sum of all the quotas is equal to the house size, so the tentative apportionments will add up to more than the house size. Critical divisors will have to be used to decide which states should receive reduced tentative apportionments.

  (b) Let state i have population p_i. Its critical divisor d_i will be the greatest divisor that will reduce its tentative apportionment n_i to $n_i - 1$. Thus, $n_i - 1 = \dfrac{p_i}{d_i}$ and hence $d_i = \dfrac{p_i}{n_i - 1}$.

  The state with the least critical divisor receives the reduced tentative apportionment, and then its critical divisor is recomputed. The process is complete when the sum of the tentative apportionments has been reduced to the house size.

  (c) The method favors small states, because it can never increase any state's tentative apportionment. The apportionments are calculated by multiplying each quota by an adjustment factor that is *less* than 1, and rounding up. A populous state's quota will be reduced more than a small state's. Also, a state will never receive more than its upper quota, but can receive less that its lower quota.

  (d) If a state's tentative apportionment is 1, then its critical divisor is ∞. Its tentative apportionment will not be reduced. Of course, this method does not work if the number of states is more than the house size!

43. (a) Let $n = \lfloor q \rfloor$. If q is between n and $n + 0.4$, then the Condorcet rounding of q is equal to n. Since $q + 0.6 < n + 1$ in this case, it is also true that $\lfloor q + 0.6 \rfloor = n$. On the other hand, if $n + 0.4 \le q < n + 1$, then the Condorcet rounding of q is $n + 1$, and also $n + 1 \le q + 0.6 < n + 1.6$, so $\lfloor q + 0.6 \rfloor = n + 1$.

(b) The method favors small states, since numbers will be rounded up more often than down; and this makes it more likely that the quotas will be adjusted downward.

(c) If the sum of the tentative apportionments is less than the house size, the critical divisor for state i, with population p_i, is the greatest divisor d_i that would apportion another seat to the state. Thus, if the tentative apportionment is n_i, then $n_i + 0.4 = \dfrac{p_i}{d_i}$, and hence $d_i = \dfrac{p_i}{n_i + 0.4}$. The state with the largest critical divisor gets the next seat, and then its critical divisor is recomputed. The process stops when the house is full.

If the total apportionment is more than the house size, then the critical divisor for state i is the least divisor that would cause the state's tentative apportionment to decrease. Thus $n_i - 1 + 0.4 = \dfrac{p_i}{d_i}$, so $d_i = \dfrac{p_i}{n_i - 0.6}$. The state with the least critical divisor of all has its tentative apportionment decreased by 1. Its critical divisor is then recomputed. The process stops when enough seats have been removed so that the number of seats apportioned is equal to the house size.

45. Let $f_i = q_i - \lfloor q_i \rfloor$ denote the fractional part of the quota for state i. Since the Hamilton method assigns to each state either its lower or its upper quota, so for each state, the absolute value $|q_i - a_i|$ is equal to either f_i (if state i received its lower quota) or $1 - f_i$ (if it received its upper quota). For convenience, let's assume that the states are ordered so that the fractions are decreasing, with f_1 the largest and f_n the smallest. If the lower quotas add up to $h - k$, where h is the house size, then states 1 through k will receive their upper quotas. The maximum absolute deviation will be the larger of $1 - f_k$ and f_{k+1}.

The maximum $|q_i - a_i|$ for the Hamilton method is less than 1, because each fractional part f_i and its complement, $1 - f_i$, is less than 1. If a particular apportionment fails to satisfy the quota condition, then for al least one state, the absolute deviation exceeds 1, and hence the maximum $|q_i - a_i|$ absolute deviation is greater than that of the Hamilton apportionment.

If an apportionment satisfies the quota condition then — as with the Hamilton method — k states receive their upper quotas and $n - k$ states receive their lower quotas.

Suppose an apportionment that satisfies the quota condition is given, and that it is not the same as the Hamilton apportionment. Let m_i denote the number of seats apportioned to state i, and let a_i be the state's apportionment under the Hamilton method. We will show that the maximum of the deviations $|q_i - m_i|$ is greater than or equal to the maximum of $|q_i - a_i|$.

Because the apportionments are different, the first k states cannot all receive their upper quotas as they did with the Hamilton method. Thus, there is a state j; where $j \leq k$, such that $m_j = \lfloor q_j \rfloor$. To compensate, some state l; where $l > k$, must get its upper quota. The absolute deviations for these states would be $|q_j - m_j| = f_j$ and $|q_l - m_l| = 1 - m_l$. But $f_j > f_k$ and $1 - f_l > 1 - f_{k+1}$ because of the way we have numbered the states, in decreasing order of fractional part.

Therefore the Hamilton apportionment, for which the greatest deviation from quota is the larger of f_k and $1 - f_{k+1}$, has the lesser maximum absolute deviation from quota.

We conclude that no apportionment is better than Hamilton's, if what we mean by "better" is "smaller maximum absolute deviation from quotas."

Chapter 15
Game Theory: The Mathematics of Competition

Exercise Solutions

1.

$$\begin{array}{cc} & \text{Row Minima} \\ \begin{bmatrix} 6 & 5 \\ 4 & 2 \end{bmatrix} & \begin{array}{c} \boxed{5} \\ 2 \end{array} \end{array}$$

Column Maxima   6 $\boxed{5}$

(a) - (b) Saddlepoint at row 1 (maximin strategy), column 2 (minimax strategy), giving value 5.

(c) Row 2 and column 1.

3.

$$\begin{array}{cc} & \text{Row Minima} \\ \begin{bmatrix} -2 & 3 \\ 1 & -2 \end{bmatrix} & \begin{array}{c} \boxed{-2} \\ \boxed{-2} \end{array} \end{array}$$

Column Maxima $\boxed{1}$ 3

(a) No saddlepoint.

(b) Rows 1 and 2 are both maximin strategies; column 1 is the minimax strategy.

(c) None.

5.

$$\begin{array}{cc} & \text{Row Minima} \\ \begin{bmatrix} -10 & -17 & -30 \\ -15 & -15 & -25 \\ -20 & -20 & -20 \end{bmatrix} & \begin{array}{c} -30 \\ -25 \\ \boxed{-20} \end{array} \end{array}$$

Column Maxima -10 -15 $\boxed{-20}$

(a) - (b) Saddlepoint at row 3 (maximin strategy), column 3 (minimax strategy), giving value -20.

(c) Column 3 dominates columns 1 and 2, so column player should avoid strategies from columns 1 and 2.

7.

		Pitcher		Row Minima
		Fastball	Knuckleball	
Batter	Fastball	0.500	0.200	$\boxed{0.200}$
	Knuckleball	0.200	0.300	$\boxed{0.200}$
	Column Maxima	0.500	$\boxed{0.300}$	

There is no saddlepoint.

		Pitcher		
		Fastball	Knuckleball	
Batter	Fastball	0.500	0.200	q
	Knuckleball	0.200	0.300	$1-q$
		p	$1-p$	

Batter:
$$E_F = 0.5q + 0.2(1-q) = 0.5q + 0.2 - 0.2q = 0.2 + 0.3q$$
$$E_K = 0.2q + 0.3(1-q) = 0.2q + 0.3 - 0.3q = 0.3 - 0.1q$$

$$E_F = E_K$$
$$0.2 + 0.3q = 0.3 - 0.1q$$
$$0.4q = 0.1$$
$$q = \frac{0.1}{0.4} = \frac{1}{4}$$
$$1 - q = 1 - \frac{1}{4} = \frac{3}{4}$$

The batter's optimal mixed strategy is $(q, 1-q) = \left(\frac{1}{4}, \frac{3}{4}\right)$.

Pitcher:
$$E_F = 0.5p + 0.2(1-p) = 0.5p + 0.2 - 0.2p = 0.2 + 0.3p$$
$$E_K = 0.2p + 0.3(1-p) = 0.2p + 0.3 - 0.3p = 0.3 - 0.1p$$

$$E_F = E_K$$
$$0.2 + 0.3p = 0.3 - 0.1p$$
$$0.4p = 0.1$$
$$p = \frac{0.1}{0.4} = \frac{1}{4}$$
$$1 - p = 1 - \frac{1}{4} = \frac{3}{4}$$

The pitcher's optimal mixed strategy is $(p, 1-p) = \left(\frac{1}{4}, \frac{3}{4}\right)$, giving value as follows.

$$E_F = E_K = E = 0.2 + 0.3\left(\tfrac{1}{4}\right) = 0.2 + 0.075 = 0.275$$

9. The following table represents the gain or loss for the businessman.

		Tax Agency		Row Minima
		Not Audit	Audit	
Businessman	Not Cheating	$100	–$100	$\boxed{-\$100}$
	Cheating	$1000	–$3000	–$3000
	Column Maxima	$1000	$\boxed{-\$100}$	

Saddlepoint is "not cheat" and "audit," giving value –$100.

11. (a)

	Officer does not patrol	Officer patrols
You park in street	0	–$40
You park in lot	–$32	–$16

(b)

	Officer does not patrol (*NP*)	Officer patrols (*P*)	
You park in street (*S*)	0	–$40	q
You park in lot (*L*)	–$32	–$16	$1-q$
	p	$1-p$	

You:
$$E_P = (0)q + (-32)(1-q) = 0 - 32 + 32q = -32 + 32q$$
$$E_{NP} = -40q + (-16)(1-q) = -40q - 16 + 16q = -16 - 24q$$
$$E_P = E_{NP}$$
$$-32 + 32q = -16 - 24q$$
$$56q = 16$$
$$q = \tfrac{16}{56} = \tfrac{2}{7} \Rightarrow 1 - q = 1 - \tfrac{2}{7} = \tfrac{5}{7}$$

Your optimal mixed strategy is $(q, 1-q) = \left(\tfrac{2}{7}, \tfrac{5}{7}\right)$.

Officer:
$$E_S = (0)p + (-40)(1-p) = 0 - 40 + 40p = -40 + 40p$$
$$E_L = -32p + (-16)(1-p) = -32p - 16 + 16p = -16 - 16p$$
$$E_S = E_L$$
$$-40 + 40p = -16 - 16p$$
$$56p = 24 \Rightarrow p = \tfrac{24}{56} = \tfrac{3}{7}$$
$$1 - p = 1 - \tfrac{3}{7} = \tfrac{4}{7}$$

The officer's optimal mixed strategy is $(p, 1-p) = \left(\tfrac{3}{7}, \tfrac{4}{7}\right)$, giving the following.

$$E_S = E_L = E = -16 - 16\left(\tfrac{3}{7}\right) \approx -16 - 6.86 = -22.86$$

The value is –$22.86.

(c) It is unlikely that the officer's payoffs are the opposite of yours—that she always benefits when you do not.

(d) Use some random device, such as a die with seven sides.

13. (a) Move first to the center box; if your opponent moves next to a corner box or to a side box, move to a corner box in the same row or column. There are now six more boxes to fill, and you have up to three more moves (if you or your opponent does not win before this point), but the rest of your strategy becomes quite complicated, involving choices like "move to block the completion of a row/column/diagonal by your opponent."

(b) Showing that your strategy is optimal involves showing that it guarantees at least a tie, no matter what choices your opponent makes.

15. Player II will choose H $\frac{1}{2}$ of the time and T $\frac{1}{2}$ of the time.

For player I, $E_H = 8\left(\frac{1}{2}\right) - 3 = 4 - 3 = 1$ and $E_T = -4\left(\frac{1}{2}\right) + 1 = -2 + 1 = -1$.

Thus, player I should always play H, winning \$1 on average.

17. Rewriting the matrix using abbreviations we have the following.

		Player II		
		F	C	R
	F	$-.25$	0	.25
Player I	BF	0	0	$-.25$
	BC	$-.25$	$-.25$	0

(a) Player I should avoid "Bet, then call" because it is dominated by "fold" (all entries in F row are bigger than corresponding entries in BC). Player II should avoid "call" because "fold" dominates it (all entries in F column are smaller than corresponding entries in C).

(b) Player I will never use "Bet, then call", and Player II will never use "Calls". Removing these, we are left with the following.

		Player II		
		F	R	
Player I	F	$-.25$	.25	q
	BF	0	$-.25$	$1-q$
		p	$1-p$	

$$E_F = -.25q + 0(1-q) = -.25q$$
$$E_R = .25q + (-.25)(1-q) = .50q - .25$$

$$E_F = E_R$$
$$-.25q = .50q - .25$$
$$-.75q = -.25$$
$$q = \frac{-.25}{-.75} = \frac{1}{3}$$
$$1 - q = 1 - \frac{1}{3} = \frac{2}{3}$$

Player I's strategy for (F, BF, BC) is $\left(\frac{1}{3}, \frac{2}{3}, 0\right)$.

Continued on next page

17. (b) continued

$$E_F = -.25p + .25(1-p) = -.25p + .25 - .25p = .25 - .50p$$

$$E_{BF} = (0)p + (-.25)(1-p) = -.25 + .25p$$

$$E_F = E_{BF}$$

$$.25 - .50p = -.25 + .25p$$

$$-.75p = -.50$$

$$p = \frac{-.50}{-.75} = \frac{2}{3}$$

$$1 - p = 1 - \frac{2}{3} = \frac{1}{3}$$

Player II's strategy for (F, C, R) is $\left(\frac{2}{3}, 0, \frac{1}{3}\right)$.

$$E_F = E_{BF} = E = -.25 + .25\left(\frac{2}{3}\right) = -\frac{1}{4} + \frac{1}{4}\left(\frac{2}{3}\right) = -\frac{3}{12} + \frac{2}{12} = -\frac{1}{12}$$

The value is value $-\frac{1}{12}$.

(c) Player II. Since the value is negative, player II's average earnings are positive and player I's are negative.

(d) Yes. Player I bets first while holding L with probability $\frac{2}{3}$. Player II raises while holding L with probability $\frac{1}{3}$, so sometimes player II raises while holding L.

19. (a) Leave umbrella at home if there is a 50% chance of rain; carry umbrella if there is a 75% chance of rain.

(b) Carry umbrella in case it rains.

(c) Saddlepoint at "carry umbrella" and "rain," giving value –2.

(d) Leave umbrella at home.

21. The Nash equilibrium outcomes are $(4,3)$ and $(3,4)$. [It would be better if the players could flip a coin to decide between $(4,3)$ and $(3,4)$.]

23. The Nash equilibrium outcome is $(2,4)$, which is the product of dominant strategies by both players.

25. The players would have no incentive to lie about the value of their own weapons unless they were sure about the preferences of their opponents and could manipulate them to their advantage. But if they do not have such information, lying could cause them to lose more than 10% of their weapons, as they value them, in any year.

27. The sophisticated outcome, x, is found as follows: Y's strategy of y is dominated; with this strategy of Y eliminated, X's strategy of x is dominated; with this strategy of X eliminated, Z's strategy of z is dominated, which is eliminated. This leaves X voting for xy (both x and y), Y voting for yz, and Z voting for zx, creating a three-way tie for x, y, and z, which X will break in favor of x.

29. The payoff matrix is as follows:

		Even		
		2	**4**	**6**
	1	(2,1)	(2,1)	(2,1)
Odd	**2**	(2,4)	(6,3)	(6,3)
	3	(2,4)	(4,8)	(10,5)

Odd will eliminate strategy 1, and Even will eliminate strategy 6, because they are dominated. In the reduced 2×2 game, Odd will eliminate strategy 5. In the reduced 1×2 game, Even will eliminate strategy 4. The resulting outcome will be $(2,4)$, in which Odd chooses strategy 3 and Even chooses strategy 2. The outcome $(2,1)$, in which Odd chooses strategy 1 and Even chooses strategy 2, is also in equilibrium.

31. If the first player shoots in the air, he will be no threat to the two other players, who will then be in a duel and shoot each other. If a second player fires in the air, then the third player will shoot one of these two, so the two who fire in the air will each have a 50–50 chance of survival. Clearly, the third player, who will definitely survive and eliminate one of her opponents, is in the best position.

33. In a duel, each player has incentive to fire – preferably first – because he or she does better whether the other player fires (leaving no survivors, which is better than being the sole victim) or does not fire (you are the sole survivor, which is better than surviving with the other player). In a truel, if you fire first, then the player not shot will kill you in turn, so nobody wants to fire first. In a four-person shoot-out, if you fire first, then you leave two survivors, who will not worry about you because you have no more bullets, leading them to duel. Thus, the incentive in a four-person shoot-out—to fire first—is the same as that in a duel.

35. Nobody will shoot.

37. The possibility of retaliation deters earlier shooting.

Chapter 16
Identification Numbers

Exercise Solutions

1. Since $3+9+5+3+8+1+6+4+0 = 48 = 9 \times 5 + 3$, the check digit is 3.

3. Since $873345672 = 7 \times 124763667 + 3$, the check digit is 3.

5. Since $30860422052 = 7 \times 4408631721 + 5$, the check digit is 5.

7. Since $3 \cdot 3 + 8 + 3 \cdot 1 + 3 + 3 \cdot 7 + 0 + 3 \cdot 0 + 9 + 3 \cdot 2 + 1 + 3 \cdot 3 = 69$, the check digit is 1.

9. Since $10 \cdot 0 + 9 \cdot 6 + 8 \cdot 6 + 7 \cdot 9 + 6 \cdot 1 + 5 \cdot 9 + 4 \cdot 4 + 3 \cdot 9 + 2 \cdot 3 = 265 = 11 \times 24 + 1$, the check digit is X.

11. The lead digits contribute $1 \cdot 9 + 3 \cdot 7 = 30$ to the weighted sum. Thus, the digit needed to make the sum evenly divisible by 10 is the same if you leave the 9 and 7 out of the calculation. (For example, if the sum were 162 including the 97 then the sum would be 132 not including them. In either case, the check digit is 8.)

13. Since $4 + 6 + 1 + 2 + 1 + 2 + 0 + 2 + 3 = 21$, the check digit is 6.

15. Since $(3 + 4 + 0 + 3 + 0 + 3 + 2 + 7) \times 2 = 44$ and one of the summands exceeds 4, we have $44 + 1 + (5 + 1 + 2 + 2 + 0 + 3 + 2 + 0) = 60$. So, the number is valid.

17. Since $7 \cdot 3 + 8 + 7 \cdot 1 + 3 + 7 \cdot 7 + 0 + 7 \cdot 0 + 9 + 7 \cdot 2 + 1 + 7 \cdot 3 = 133$, the check digit is 7. This check-digit scheme will detect all single-digit errors.

19. In the odd-numbered positions, if a digit a is replaced by the digit b where $a - b$ is even, the error is not detected. This occurs if a and b are both even or both odd.

21. We begin with $(3 + 0 + 2 + 6 + 0 + 9 + 4 + 1) \times 2 = 50$. Adding 2, we obtain 52 and have the following.
$$52 + 0 + 1 + 5 + 0 + 1 + 6 + 3 = 68$$
So, the check digit is 2.

23. (a) Since $1 \cdot 0 + 1 \cdot 1 + 3 \cdot 2 + 3 \cdot 1 + 1 \cdot 6 + 3 \cdot 9 + 1 \cdot 0 = 43$, the check digit is 7.

    (b) Since $1 \cdot 0 + 1 \cdot 2 + 3 \cdot 7 + 3 \cdot 4 + 1 \cdot 5 + 3 \cdot 5 + 1 \cdot 1 = 56$, the check digit is 4.

    (c) Since $1 \cdot 0 + 1 \cdot 7 + 3 \cdot 6 + 3 \cdot 0 + 1 \cdot 0 + 3 \cdot 2 + 1 \cdot 2 = 33$, the check digit is 7.

    (d) Since $1 \cdot 0 + 1 \cdot 4 + 3 \cdot 9 + 3 \cdot 6 + 1 \cdot 5 + 3 \cdot 8 + 1 \cdot 0 = 78$, the check digit is 2.

25. First observe that the given number 0-669-03925-4 results in a weighted sum that has a remainder of 5 after division by 11. So all we need to do is check for successive pairs of digits of this number that results in a contribution to the weighted sum of 5 less or 6 more, since either of these will make the weighted sum divisible by 11. Checking each pair of consecutive digits, we see that 39 contributes $5 \cdot 3 + 4 \cdot 9 = 51$ whereas 93 contributes $5 \cdot 9 + 4 \cdot 3 = 57$. So, the correct number is 0-669-09325-4.

27. Notice that when we add the weighted sum used for the actual check digit:
$$7a_1 + 3a_2 + 9a_3 + 7a_4 + 3a_5 + 9a_6 + 7a_7 + 3a_8$$
and the weighted sum
$$3a_1 + 7a_2 + a_3 + 3a_4 + 7a_5 + a_6 + 3a_7 + 7a_8,$$
we obtain
$$10a_1 + 10a_2 + 10a_3 + 10a_4 + 10a_5 + 10a_6 + 10a_7 + 10a_8,$$
which always ends with 0. So, the actual check digit and the check digit calculated with the weighted sum $3a_1 + 7a_2 + a_3 + 3a_4 + 7a_5 + a_6 + 3a_7 + 7a_8$ are both 0 or their sum is 10.

29. The mistake of reading a 2 as a 7 is detected because the sum of the digits would be odd. The mistake of reading a 2 as an 8 is not detected because the sum of the digits would remain even. An error is detected when an odd digit is misread as an even one or vice versa because the sum of the digits changes from even to odd or odd to even.

31. The computer need not know which digit is the check digit since it merely checks to see if the weighted sum is divisible by 9 for traveler's checks and divisible by 10 for the other two.

33. Since the remainder after dividing by 9 is less than 9, 9 cannot be a check digit.

35. Because the remainder after dividing by 7 is less than 7, the digits 7, 8, and 9 cannot be a check digit.

37. Yes. The ISBN-10 scheme detects all transposition errors.

39. For the transposition to go undetected, it must be the case that the difference of the correct number and the incorrect number is evenly divisible by 11. That is,
$$(10a_1 + 9a_2 + 8a_3 + \cdots + a_{10}) - (10a_3 + 9a_2 + 8a_1 + \cdots + a_{10})$$
is divisible by 11. This reduces to $2a_1 - 2a_3 = 2(a_1 - a_3)$ is divisible by 11. But $2(a_1 - a_3)$ is divisible by 11 only when $a_1 - a_3$ is divisible by 11 and this only happens when $a_1 - a_3 = 0$. In this case, there is no error. The same argument works for the fourth and sixth digits.

41. The combination 72 contributes $7 \cdot 1 + 2 \cdot 3 = 13$ or $7 \cdot 3 + 2 \cdot 1 = 23$ (depending on the location of the combination) towards the total sum, while the combination 27 contributes $2 \cdot 1 + 7 \cdot 3 = 23$ or $2 \cdot 3 + 7 \cdot 1 = 13$. So, the total sum resulting from the number with the transposition is still divisible by 10. Therefore, the error is not detected. When the combination 26 contributes $2 \cdot 1 + 6 \cdot 3 = 20$ towards the total sum, the combination 62 contributes $6 \cdot 1 + 2 \cdot 3 = 12$ toward the total sum; so the new sum will not be divisible by 10. Similarly, when the combination 26 contributes $2 \cdot 3 + 6 \cdot 1 = 12$ to the total, the combination 62 contributes $6 \cdot 3 + 2 \cdot 1 = 20$ to the total. So, the total for the number resulting from the transposition will not be divisible by 10 and the error is detected. In general, an error that occurs by transposing ab to ba is undetected if and only if $a - b = \pm 5$.

43. The error $\cdots abc \cdots \rightarrow \cdots cba \cdots$ is undetectable if and only if $a - c = \pm 5$. To see this in the case that the weights for abc are 7, 3, 9, notice that a and c contribute $7a + 9c$ toward the weighted sum, whereas in the case of cba, the c and a contribute $7c + 9a$. Thus, the error is undetectable if and only if $7a + 9c$ and $7c + 9a$ contribute equal amounts to the last digit of the weighted sum. This means that they differ by a multiple of 10. That is, $-2a + 2c = 2(c - a)$ is a multiple of 10. This occurs when $c - a = 0$ or $c - a = \pm 5$. When $c - a = 0$, there is no error.

45. There are a few situations where an error is not detected.

- Since any error in the position with weight 10 does not change the last digit of the weighted sum, no error in that position is detected. Thus, all errors in position 3 are undetectable.
- In the position with weight 5, replacing an even digit by any other even digit is not detected. So if b were to replace a in that position where $b-a$ is an even integer, the error would not be detected. This occurs in position 8.
- In positions with weights 12, 8, 6, 4, or 2, replacing b by a is undetectable if $b-a = \pm 5$. These are positions 1, 5, 7, 9, and 11.

47. Since both numbers are valid the difference of the weighted sums is divisible by 10. That is, $(7w+3+2w+1+5w+6+7w+4)-(7w+3+2w+1+5w+6+6w+1)$ is divisible by 10. The difference simplifies to $w+3$. So, $w=7$.

49. **(a)** The code is 51593-2067; since $5+1+5+9+3+2+0+6+7=38$, the check digit is 2.

(b) The code is 50347-0055; since $5+0+3+4+7+0+0+5+5=29$, the check digit is 1.

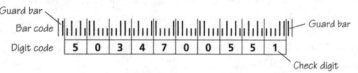

(c) The code is 44138-9901; since $4+4+1+3+8+9+9+0+1=39$, the check digit is 1.

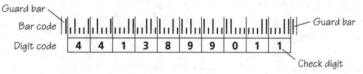

51. **(a)** Since the sixth block of five bars (ignoring the first bar) has one long bar and four short bars, that block is incorrect. Call the digit corresponding to that block x. Then the code is $20782x960$. Since the sum of the digits is $x+41$, $x=9$. Finally, we write 20782-9960.

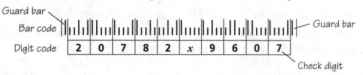

(b) Since the eighth block of five bars (ignoring the first bar) has three long bars and two short bars, that block is incorrect. Call the digit corresponding to that block x. Then the code is $5543599x2$. Since the sum of the digits is $x+42$, $x=8$. Finally, we write 55435-9982.

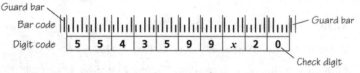

(c) Since the tenth block of five bars (ignoring the first bar) has one long bar and four short bars, that block is incorrect. Since this is the check digit, the nine-digit ZIP code is 52735-2101.

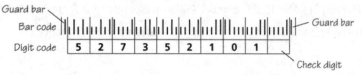

53. If a double error in a block results in a new block that does not contain exactly two long bars, we know this block has been misread. If a double error in a block of five results in a new block with exactly two long bars, the block now gives a different digit from the original one. If no other digit is in error, the check digit catches the error, since the sum of the 10 digits will not end in 0. So, in every case an error has been detected. Errors of the first type can be corrected just as in the case of a single error. When a double error results in a legitimate code number, there is no way to determine which digit is incorrect.

55. The strings are *aaabb*, *aabab*, *aabba*, *abaab*, *ababa*, *abbaa*, *baaab*, *baaba*, *babaa*, and *bbaaa* (in alphabetical order). If you replace each short bar in the bar code table (page 517) by an *a* and replace each long bar in the bar code table by a *b*, the resulting strings are listed in alphabetical order.

57. Since there is an even number of 1's in 1000100, the scanner is reading from right to left.

59. Wyoming, Nevada, and Alaska.

61. The size of the population.

63. The Canadian scheme detects any transposition error involving adjacent characters. Also, there are $26^3 \times 10^3 = 17,576,000$ possible Canadian codes but only $10^5 = 100,000$ U.S. five-digit ZIP codes. Hence the Canadian scheme can target a location more precisely.

65. Skow $\to$ Sko $\to$ 220 $\to$ 20 $\to$ 0 $\to$ all numbers gone $\to$ S-000
Sachs $\to$ Sacs $\to$ 2022 $\to$ 202 $\to$ 02 $\to$ 2 $\to$ S-200
Lennon $\to$ Lennon $\to$ 405505 $\to$ 0505 $\to$ 55 $\to$ L-550
Lloyd $\to$ Lloyd $\to$ 44003 $\to$ 403 $\to$ 03 $\to$ 3 $\to$ L-300
Ehrheart $\to$ Ereart $\to$ 060063 $\to$ 06063 $\to$ 6063 $\to$ 663 $\to$ E-663
Ollenburger $\to$ Ollenburger $\to$ 04405106206 $\to$ 0405106206 $\to$ 405106206 $\to$ 451626 $\to$ O-451

67. A person born in 1999 is too young for a driver's license.

69. For a woman born in November or December the formula $40(m-1)+b+600$ gives a number requiring four digits.

71. The 58 indicates that the year of birth is 1958. Since 818 is larger than $12 \cdot 31 = 372$, we subtract 600 from 818 to obtain 218. Then $218 = 7 \cdot 31 + 1$ tells us that the person was born on the first day of the eighth month. So, the birth date is August 1, 1958.

73. The Soundex code for Gallihan is as follows.

Gallihan $\to$ Gallian $\to$ 2044005 $\to$ 20405 $\to$ 0405 $\to$ 45 $\to$ G-450

The following share the same Soundex code. They all coincide with Gallihan on the third step and thereafter.

Gallian $\to$ Gallian $\to$ 2044005 $\to$ 20405 $\to$ 0405 $\to$ 45 $\to$ G-450

Galliam $\to$ Galliam $\to$ 2044005 $\to$ 20405 $\to$ 0405 $\to$ 45 $\to$ G-450

Gilliam $\to$ Gilliam $\to$ 2044005 $\to$ 20405 $\to$ 0405 $\to$ 45 $\to$ G-450

Galin $\to$ Galin $\to$ 20405 $\to$ 20405 $\to$ 0405 $\to$ 45 $\to$ G-450

75. Since $248 = 63(3)+58+1$, the number 248 corresponds to a female born on March 29; since $601 = 63(9)+34$ the number corresponds to a male born on September 17.

77. Likely circumstances could be twins; sons named after their fathers (such as John L. Smith, Jr.); common names such as John Smith and Mary Johnson; and states that do not include year of birth in the code.

79. Because of the short names and large population there would be a significant percentage of people whose names would be coded the same.

Chapter 17
Information Science

Exercise Solutions

1. For 0101:

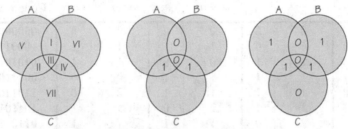

The code word given by regions I, II, III, IV, V, VI, VII is 0101110.

For 1011:

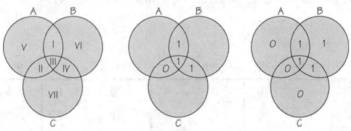

The code word given by regions I, II, III, IV, V, VI, VII is 1011010.

For 1111:

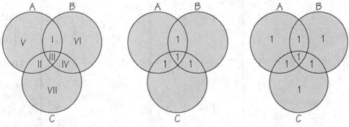

The code word given by regions I, II, III, IV, V, VI, VII is 1111111.

3. **(a)** When you compare 1**1011011** and 1**0100110**, they differ by 6 digits. Thus, the distance is 6.

   (b) When you compare **01110**100 and **11101**100, they differ by 3 digits. Thus, the distance is 3.

5. There would be no change to 1001101 since the total number of 1's in each circle is even.

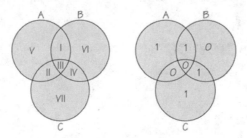

7. Consider the following table.

	$a_2 + a_3$	c_1	$a_1 + a_3$	c_2	$a_1 + a_2$	c_3	Code word
000	0	0	0	0	0	0	000000
100	0	0	1	1	1	1	100011
010	1	1	0	0	1	1	010101
001	1	1	1	1	0	0	001110
110	1	1	1	1	2	0	110110
101	1	1	2	0	1	1	101101
011	2	0	1	1	1	1	011011
111	2	0	2	0	2	0	111000

Thus, the binary linear code is 000000, 100011, 010101, 001110, 110110, 101101, 011011, 111000.

9. Consider the following table.

	$a_2 + a_3 + a_4$	c_1	$a_2 + a_4$	c_2	$a_1 + a_2 + a_3$	c_3	Code word
0000	0	0	0	0	0	0	0000000
1000	0	0	0	0	1	1	1000001
0100	1	1	1	1	1	1	0100111
0010	1	1	0	0	1	1	0010101
0001	1	1	1	1	0	0	0001110
1100	1	1	1	1	2	0	1100110
1010	1	1	0	0	2	0	1010100
1001	1	1	1	1	1	1	1001111
0110	2	0	1	1	2	0	0110010
0101	2	0	2	0	1	1	0101001
0011	2	0	1	1	1	1	0011011
1110	2	0	1	1	3	1	1110011
1101	2	0	2	0	2	0	1101000
1011	2	0	1	1	2	0	1011010
0111	3	1	2	0	2	0	0111100
1111	3	1	2	0	3	1	1111101

Thus, the binary linear code is 0000000, 1000001, 0100111, 0010101, 0001110, 1100110, 1010100, 1001111, 0110010, 0101001, 0011011, 1110011, 1101000, 1011010, 0111100, 1111101. No, since 1000001 has weight 2.

11. Consider the following table.

	$a_1 + a_2$	c_1	$a_2 + a_3$	c_2	$a_1 + a_3$	c_3	Code word
000	0	**0**	0	**0**	0	**0**	**000000**
100	1	**1**	0	**0**	1	**1**	**100101**
010	1	**1**	1	**1**	0	**0**	**010110**
001	0	**0**	1	**1**	1	**1**	**001011**
110	2	**0**	1	**1**	1	**1**	**110011**
101	1	**1**	1	**1**	2	**0**	**101110**
011	1	**1**	2	**0**	1	**1**	**011101**
111	2	**0**	2	**0**	2	**0**	**111000**

Thus, the binary linear code is 000000, 100101, 010110, 001011, 110011, 101110, 011101, 111000. 001001 is decoded as 001011; 011000 is decoded as 111000; 000110 is decoded as 010110; 100001 is decoded as 100101.

13. Consider the following table.

	Weight	Append	Code word
0000000	0	**0**	**00000000**
0001011	3	**1**	**00010111**
0010111	4	**0**	**00101110**
0100101	3	**1**	**01001011**
1000110	3	**1**	**10001101**
1100011	4	**0**	**11000110**
1010001	3	**1**	**10100011**
1001101	4	**0**	**10011010**
0110010	3	**1**	**01100101**
0101110	4	**0**	**01011100**
0011100	3	**1**	**00111001**
1110100	4	**0**	**11101000**
1101000	3	**1**	**11010001**
1011010	4	**0**	**10110100**
0111001	4	**0**	**01110010**
1111111	7	**1**	**11111111**

The extended code is 00000000, 00010111, 00101110, 01001011, 10001101, 11000110, 10100011, 10011010, 01100101, 01011100, 00111001, 11101000, 11010001, 10110100, 01110010, 11111111. The code will detect any three errors or correct any single error.

15. There are $2^5 = 32$ possible messages of length 5. There are $2^8 = 256$ possible received words.

17. Consider the following table.

	$a_1 + a_2$	$c_1 = (a_1 + a_2) \bmod 3$	$2a_1 + a_2$	$c_2 = (2a_1 + a_2) \bmod 3$	Code word
00	0	**0**	0	**0**	**0000**
10	1	**1**	2	**2**	**1012**
20	2	**2**	4	**1**	**2021**
01	1	**1**	1	**1**	**0111**
02	2	**2**	2	**2**	**0222**
11	2	**2**	3	**0**	**1120**
22	4	**1**	6	**0**	**2210**
21	3	**0**	5	**2**	**2102**
12	3	**0**	4	**1**	**1201**

Thus, the ternary code is 0000, 1012, 2021, 0111, 0222, 1120, 2210, 2102, 1201.

19. There are $3^4 = 81$ possible messages of length 5. There are $3^6 = 729$ possible received words.

21. 001100001111000 can be written as 0, 0, 110, 0, 0, 0, 111, 10, 0, 0. Thus we have, AATAAAGCAA.

23. 1111, 0, 10, 0, 0, 1110, 0, 10, 10 would be written as 111101000111001010. 001000110011110111010 can be written as 0, 0, 10, 0, 0, 110, 0, 1111, 0, 1110, 10. Thus we have *AABAACAEADB*.

25. *t*, *n*, and *r* would be the most frequently occurring consonants. *e* would be the most frequently occurring vowel.

27. In the Morse code, a space is needed to determine where each code word ends. In a fixed-length code of length k, a word ends after each k digits.

29. Given the following:

13403	13403
$13739 - 13403$	336
$13816 - 13739$	77
$13767 - 13816$	−49
$13820 - 13767$	53
$13759 - 13820$	−61
$13779 - 13759$	20
$13878 - 13779$	99
$13913 - 13878$	35
$13896 - 13913$	−17

the compressed numbers are 13403 336 77 −49 53 −61 20 99 35 −17; We go from 50 characters to 27 characters, a reduction of $\dfrac{50-27}{50} = \dfrac{23}{50} = 46\%$.

31. 11100100001001110110011010 can be written as 1110, 01, 00, 00, 10, 01, 110, 110, 01, 10, 10. Thus, the message is decoded to be *BCEEFCDDCFF*.

33. *B* is the least likely letter, *J* is the second least likely, and *G* is the third least likely letter.

35. RETREAT would be encrypted as UHWUHDW.   DGYDQFH would be decrypted as ADVANCE.

37. One can take a position, such as 0 and determine that it takes 13 iterations to arrive at 0 again.

$$(0+8) \bmod 26 = 8 \bmod 26 = 8$$

$$(8+8) \bmod 26 = 16 \bmod 26 = 16$$

$$(16+8) \bmod 26 = 24 \bmod 26 = 24$$

$$(24+8) \bmod 26 = 32 \bmod 26 = 6$$

$$(6+8) \bmod 26 = 14 \bmod 26 = 14$$

$$(14+8) \bmod 26 = 22 \bmod 26 = 22$$

$$(22+8) \bmod 26 = 30 \bmod 26 = 4$$

$$(4+8) \bmod 26 = 12 \bmod 26 = 12$$

$$(12+8) \bmod 26 = 20 \bmod 26 = 20$$

$$(20+8) \bmod 26 = 28 \bmod 26 = 2$$

$$(2+8) \bmod 26 = 10 \bmod 26 = 10$$

$$(10+8) \bmod 26 = 18 \bmod 26 = 18$$

$$(18+8) \bmod 26 = 26 \bmod 26 = 0$$

39. Given RETREAT to encrypt with key 5 we have the following.

Decrypted	Location	×5			Encrypted
R	17	85	$7 = 85 \bmod 26$	7	H
E	4	20	$20 = 20 \bmod 26$	20	U
T	19	95	$17 = 95 \bmod 26$	17	R
R	17	85	$7 = 85 \bmod 26$	7	H
E	4	20	$20 = 20 \bmod 26$	20	U
A	0	0	$0 = 0 \bmod 26$	0	A
T	19	95	$17 = 95 \bmod 26$	17	R

The encrypted message is HURHUAR.

41. There is no integer j such that $2j = 1 \bmod 26$.

43. Given the key word BEATLES, we note that B is in the position 1; E is in position 4; A is in position 0; T is in position 19; L is in position 11; S is in position 18.

Encrypted	Location				Decrypted
S	18	1	$18 = 18 \bmod 26 = (17+1) \bmod 26$	17	R
S	18	4	$18 = 18 \bmod 26 = (14+4) \bmod 26$	14	O
L	11	0	$11 = 11 \bmod 26 = (11+0) \bmod 26$	11	L
E	4	19	$4 = 30 \bmod 26 = (11+19) \bmod 26$	11	L
T	19	11	$19 = 19 \bmod 26 = (8+11) \bmod 26$	8	I
R	17	4	$17 = 17 \bmod 26 = (13+4) \bmod 26$	13	N
Y	24	18	$24 = 24 \bmod 26 = (6+18) \bmod 26$	6	G
T	19	1	$19 = 19 \bmod 26 = (18+1) \bmod 26$	18	S
X	23	4	$23 = 23 \bmod 26 = (19+4) \bmod 26$	19	T
O	14	0	$14 = 14 \bmod 26 = (14+0) \bmod 26$	14	O
G	6	19	$6 = 32 \bmod 26 = (13+19) \bmod 26$	13	N
P	15	11	$15 = 15 \bmod 26 = (4+11) \bmod 26$	4	E
W	22	4	$22 = 22 \bmod 26 = (18+4) \bmod 26$	18	S

The original message was ROLLING STONES.

45. (a)   10111011
      + 01111011
      11000000

(b)   11101000
      + 01110001
      10011001

47. 1.   VIP converts to 220916.

2.   We will send 22, 09, and 16 individually.

3.   Since $p = 5$ and $q = 17$, $n = pq = 5 \cdot 17 = 85$.

Since $\text{GCF}(22, 85) = 1$, $\text{GCF}(9, 85) = 1$, and $\text{GCF}(16, 85) = 1$, we can proceed. (GCF stands for *greatest common factor*.) Thus, $M_1 = 22$, $M_2 = 09$, and $M_3 = 16$.

4.   Since $R_i = M_i^r \bmod n$ and $r = 3$, we have the following.

$$R_1 = 22^3 \bmod 85 = (2 \cdot 11)^3 \bmod 85 = \left[(2^3 \cdot 11) \cdot 11^2 \right] \bmod 85$$

$$= \left[\left[(2^3 \cdot 11) \bmod 85 \right] \left[11^2 \bmod 85 \right] \right] \bmod 85$$

$$= \left[(88 \bmod 85)(121 \bmod 85) \right] \bmod 85 = (3 \cdot 36) \bmod 85 = 108 \bmod 85 = 23$$

$$R_2 = 9^3 \bmod 85 = (3^2)^3 \bmod 85 = 3^6 \bmod 85 = \left[(3^5 \bmod 85)(3 \bmod 85) \right] \bmod 85$$

$$= \left[(243 \bmod 85)(3 \bmod 85) \right] \bmod 85 = (73 \cdot 3) \bmod 85 = 219 \bmod 85 = 49$$

$$R_3 = 16^3 \bmod 85 = (2^4)^3 \bmod 85 = 2^{12} \bmod 85 = \left[(2^7 \bmod 85)(2^5 \bmod 85) \right] \bmod 85$$

$$= \left[(128 \bmod 85)(32 \bmod 85) \right] \bmod 85 = (43 \cdot 32) \bmod 85 = (43 \cdot 2 \cdot 16) \bmod 85$$

$$= \left[(43 \cdot 2) \cdot 16 \right] \bmod 85 = \left[(86 \bmod 85)(16 \bmod 85) \right] \bmod 85$$

$$= \left[1 \cdot (16 \bmod 85) \right] \bmod 85 = (16 \bmod 85) = 16$$

Thus, the numbers sent are 23, 49, 16.

49. 1.   Since $p = 5$ and $q = 17$, $n = pq = 5 \cdot 17 = 85$.

2.   Since $p - 1 = 5 - 1 = 4$ and $q - 1 = 17 - 1 = 16$, m will be the least common multiple of 4 and 16, namely 16.

3.   We need to choose r such that it has no common divisors with 16. Thus, r can be 5. This confirms that $r = 5$ is a valid choice.

4.   We need to find s such that $rs = 1 \bmod m$.

$$r^2 \bmod m = 5^2 \bmod 16 = 25 \bmod 16 = 9$$
$$r^3 \bmod m = 5^3 \bmod 16 = 125 \bmod 16 = 13$$
$$r^4 \bmod m = 5^4 \bmod 16 = 625 \bmod 16 = 1$$

Thus $t = 4$.

Since $s = r^{t-1} \bmod m$, where $r = 5$ and $t = 4$, we have the following.

$$s = 5^{4-1} \bmod 16 = 5^3 \bmod 16 = 125 \bmod 16 = 13$$

51. N converts to 14 and O converts to 15, but 14 and 77 have a greatest common divisor of 7. On the other hand, using blocks of length 4, NO converts to 1415 and the greatest common divisor of 77 and 1415 is 1.

53. As the following shows, since the entries in the column for the variable P are exactly the same as the entries in the column for $P \vee (P \wedge Q)$, the two expressions are logically equivalent.

P	Q	$P \wedge Q$	$P \vee (P \wedge Q)$
T	T	T	T
T	F	F	T
F	T	F	F
F	F	F	F

55. First we construct the truth table for $\neg(P \wedge Q)$.

P	Q	$P \wedge Q$	$\neg(P \wedge Q)$
T	T	T	F
T	F	F	T
F	T	F	T
F	F	F	T

Next we construct the truth table for $\neg P \vee \neg Q$

P	Q	$\neg P$	$\neg Q$	$\neg P \vee \neg Q$
T	T	F	F	F
T	F	F	T	T
F	T	T	F	T
F	F	T	T	T

Since the last columns of the two truth tables are identical, we have shown that $\neg(P \wedge Q)$ is logically equivalent to $\neg P \vee \neg Q$.

57. First we construct the truth table for $P \wedge (Q \vee R)$.

P	Q	R	$Q \vee R$	$P \wedge (Q \vee R)$
T	T	T	T	T
T	T	F	T	T
T	F	T	T	T
T	F	F	F	F
F	T	T	T	F
F	T	F	T	F
F	F	T	T	F
F	F	F	F	F

Next we construct the truth table for $(P \wedge Q) \vee (P \wedge R)$.

P	Q	R	$P \wedge Q$	$P \wedge R$	$(P \wedge Q) \vee (P \wedge R)$
T	T	T	T	T	T
T	T	F	T	F	T
T	F	T	F	T	T
T	F	F	F	F	F
F	T	T	F	F	F
F	T	F	F	F	F
F	F	T	F	F	F
F	F	F	F	F	F

Since the last columns of the two truth tables are identical, the expression $P \wedge (Q \vee R)$ is logically equivalent to $(P \wedge Q) \vee (P \wedge R)$.

59. The truth table for $\neg P \vee Q$ is as follows.

P	Q	$\neg P$	$\neg P \vee Q$
T	T	F	T
T	F	F	F
F	T	T	T
F	F	T	T

Because the last column of this truth table is identical for the one for $P \rightarrow Q$, we conclude that the two expressions are logically equivalent.

61. Let P denote "it snows" and let Q denote "there is school." Then the statement "If it snows, there will be no school" can be expressed as $P \to \neg Q$. Similarly, the statement "it is not the case that: it snows and there is school" can be expressed as $\neg(P \wedge Q)$. We now construct the truth tables for each of these expressions. A truth table for $P \to \neg Q$ is as follows.

P	Q	$\neg Q$	$P \to \neg Q$
T	T	F	F
T	F	T	T
F	T	F	T
F	F	T	T

A truth table for $\neg(P \wedge Q)$ is as follows.

P	Q	$P \wedge Q$	$\neg(P \wedge Q)$
T	T	T	F
T	F	F	T
F	T	F	T
F	F	F	T

Because the two tables have identical last columns, the two expressions are logically equivalent.

63. **(a)**

s	11110001
t	00101110
$s \wedge t$	00100000

(b)

s	01110001
t	10111110
$s \wedge t$	00110000

65. $s \wedge 11100000 = t \wedge 11100000$ is the same as $s \wedge 11100000 - t \wedge 11100000 = 00000000$ but in mod 2, subtraction is the same as addition since $1 + 1 = 0$.

67. s has a 0 in position 2, 4, and 6 and a 1 in position 8.

69. Since $s \wedge 11100111 = 01100010$, we know that s must be of the form $011xy010$ where x and y represent unknown binary digits. Since two binary digits are not known, there are $2^2 = 4$ possible binary strings that satisfy the relation. These binary strings are 01100010, 01101010, 01110010, and 01111010.

71. The 255 in the subnet mask means the bits in 8 have to match exactly. So the network address is 8.0.0.0.

73. No. Since 172.16.17.30 can be expressed as 10101100.00010000.00010001.00011110 and 255.255.255.240 can be expressed as 11111111.11111111.11111111.11110000, we have the following.

10101100.00010000.00010001.00011110

$\wedge$ 11111111.11111111.11111111.11110000

10101100.00010000.00010001.00010000

The network address for 172.16.17.30 with subnet mask 255.255.255.240 is therefore 172.16.17.16.

Since 172.16.17.15 can be expressed as 10101100.00010000.00010001.00001111 and 255.255.255.240 can be expressed as 11111111.11111111.11111111.11110000, we have the following.

10101100.00010000.00010001.00001111

$\wedge$ 11111111.11111111.11111111.11110000

10101100.00010000.00010001.00000000

The network address for 172.16.17.15 with subnet mask 255.255.255.240 is therefore 172.16.17.0.

Chapter 18
Growth and Form

Exercise Solutions

1. **(a)** none

   (b) $4 \text{ in.} \times \frac{4}{3} = \frac{16}{3} \text{ in.} = 5\frac{1}{3} \text{ in.}$

   (c) $6 \text{ in.} \times \frac{3}{4} = \frac{18}{4} \text{ in.} = \frac{9}{2} \text{ in.} = 4\frac{1}{2} \text{ in.}$

3. **(a)** The linear scaling factor is $\dfrac{4 \text{ cm}}{160 \text{ cm}} = \dfrac{1}{40} = 0.025$.

   (b) The volume of the real person goes up as the cube of the scaling factor and so is $40^3 = 64,000$ times as large as the volume of the Lego.

   (c) $40 \times 10 \text{ cm} = 400 \text{ cm} = 4.00 \text{m} = 400 \text{ cm} \times \dfrac{1 \text{ in}}{2.54 \text{ cm}} \approx 157.5 \text{ in.} \approx 13.1 \text{ ft.}$

5. The linear scaling factor for men compared to women (on average) is 1.08; if the brain scales as the cube of height, then men's brains (on average) would be $1.08^3 \approx 1.26$ times as large as women's, or 26% larger.

7. **(a)** The new altar would have a volume 8 times as large – not "8 times greater than" or "8 times larger than", and definitely not twice as large, as the old altar.

   (b) Since the volume scales as the cube of the side, the side scales as the cube root of the volume: $\sqrt[3]{2} \approx 1.26$.

9. The writer uses both multiplicative and additive language together. Better to say: "Safari takes only half as much time as Internet Explorer to load a page, and about 60% as long as Firefox 2."

11. Nothing can decrease 150% without becoming negative. It's not clear what the writer meant: that the stock has fallen to two-thirds of its former price, or maybe to 40% of its former price.

13. Answers will vary.

15. $\$0.90 = \$0.90 \times \dfrac{€1}{\$1.47} = €\dfrac{0.90}{\$1.47} \approx €0.61$.

17. The car gets 100 km per 7.3 L, which is $\dfrac{100 \text{ km}}{7.3 \text{ L}} = \dfrac{100 \times 0.621 \text{ mi}}{7.3 \times 0.2642 \text{gal}} \approx 32.2$ or 32 mpg.

19. (a) $\left(\dfrac{1}{87}\right)^3 \times 88$ tons ≈ 0.00013364 tons.

(b) We assume that all parts of the scale model are made of the same materials as the real locomotive.

(c) 0.00013364 tons $= 0.00013364 \times 2000$ lb ≈ 0.267 lb.

(d) 0.267 lb $= 0.267 \times 0.45359237$ kg ≈ 0.121 kg.

(e) 0.121 kg $= 0.121$ kg $\times \dfrac{1 \text{ metric tonne}}{1000 \text{ kg}} = 0.000121$ metric tonnes.

21. $\dfrac{€1.42}{1\,\text{L}} \times \dfrac{\$1.25}{€1} \times \dfrac{1\,\text{L}}{1000\,\text{cm}^3} \times \dfrac{(2.54\,\text{cm})^3}{(1\,\text{in})^3} \times \dfrac{231\,\text{in}^3}{1\,\text{gal}} \approx \$6.72/\text{gal}.$

23. $\dfrac{620\,\text{m}}{364.4\,\text{smoots}} \times \dfrac{100\,\text{cm}}{1\,\text{m}} \times \dfrac{1\,\text{in}}{2.54\,\text{cm}} \approx 67.0$ in./smoot or about $\left(5\text{ ft }7\text{ in.}\right)$/smoot.

25. (a) $55\text{ ft/s} = \dfrac{55 \times 1\,\text{ft}}{1\,\text{s}} = \dfrac{55 \times \dfrac{1}{5280}\,\text{mi}}{\dfrac{1}{3600}\,\text{h}} = \dfrac{55 \times 3600\,\text{mi}}{5280\,\text{h}} = 37.5$ mph.

(b) $1\text{ ft/s} = \dfrac{\dfrac{1}{5280}\,\text{mi}}{\dfrac{1}{3600}\,\text{h}} = \dfrac{3600}{5280}\,\text{mph} \approx 0.68182$ mph, so $41\text{ ft/s} \approx 41 \times 0.68182$ mph ≈ 28.0 mph.

(c) $\dfrac{1\,\text{mi}}{0.5\,\text{min}} = \dfrac{1\,\text{mi}}{0.5\,\text{min}} \times \dfrac{60\,\text{min}}{1\,\text{h}} = 120$ mph.

27. (a) $\dfrac{1.00\,\text{Middie} - 0.50\,\text{Middie}}{1.00\,\text{Middie}} \times 100\% = 50\%.$

(b) $\dfrac{1.00\,\text{Middie} - 0.50\,\text{Middie}}{0.50\,\text{Middie}} \times 100\% = 100\%.$

29. (a) $\dfrac{\$1.442 - \$0.862}{\$1.442} \times 100\% \approx 40\%.$

(b) $\dfrac{\$1.442 - \$0.862}{\$0.862} \times 100\% \approx 67\%.$

31. (a) For Option A, yes; for Option B, no. Let the previous value of the currency be C and the new value be D. Then Option A gives $\dfrac{D-C}{D}\times 100\% = \left(1-\dfrac{C}{D}\right)\times 100\% < -100\%$ if $C > 2D$. Option B gives $\dfrac{D-C}{C}\times 100\% = \left(\dfrac{D}{C}-1\right)\times 100\% \geq -100\%$ for all nonzero C, D.

(b) If the new trading value is higher than the old one, the percentage in Option B is higher than that in A: With $D > C$, Option A $= \dfrac{D-C}{D}\times 100\% < \dfrac{D-C}{C}\times 100\% =$ Option B. If the new trading value is lower than the old one, then both options give negative numbers but the absolute value of the percentage in Option B is higher than that in A: With $D < C$, Option A $\dfrac{D-C}{D}\times 100\% < \dfrac{D-C}{C}\times 100\% =$ Option B. In both cases, the absolute value of the percentage is higher for Option B.

(c) Either way, use Option B.

33. (a) The layer of soil has volume 100 ft$^2 \times 0.5$ ft $= 50$ ft^3. The density is the weight divided by the volume, so $45{,}000$ lb/50 ft$^3 = 900$ lb/ft^3.

(b) The density of steel is 500 lb/ft^3, so the claim is that the soil is $900/500 \approx 1.8 \approx 2$ times as dense as steel.

(c) Since 230 lb of compost is supposed to add about 5%, the original should be about 230 lb divided by 0.05, or $4{,}600$ lb. The revised quotation should say that the mineral soil weighs about $4{,}500$ to $4{,}600$ lb.

35. (a) Weight scales as the cube of the linear scaling factor, so KK would have to weigh 400 lb $\times 10^3 = 400{,}000$ lb.

(b) The surface area of feet scales as the square of the linear scaling factor, so the area of KK's feet is 1 ft$\times 10^2 = 100$ ft^2, and the pressure is $400{,}000$ lb/100 ft$^2 = 4{,}000$ lb/ft$^2 = 4{,}000$ lb/144 in.$^2 \approx 28$ lb/in.^{2}.

37. We have $r = 3$ ft and $h = 3.5$ ft, so the tub has volume $\pi r^2 h \approx (3.14)(3^2)(3.5)$ ft$^3 \approx 99.0$ ft^3. Per the text, 1 ft^3 of water weighs about 62 lb, so the water in the spa weighs 99.0 ft$^3 \times 62$ lb/ft$^3 \approx 6100$ lb ≈ 2800 kg.

39. The lights are strung around the outside of the tree branches, so in effect they cover the outside "area" of the tree (thought of as a cone). Hence, the number of strings needed grows in proportion to the square of the height: a 30-ft tree will need $5^2 = 25$ times as many strings as a 6-ft tree. However, you could also argue that a 30-ft tree is meant to be viewed from farther away, so that stringing the lights farther apart on the 30-ft tree would produce the same effect as with the shorter tree.

41. The lower estimate of 17 in. for a cubit leads to a height of 6×17 in.$+ 9$ in. $= 111$ in. $= 9$ ft 3 in. $= 111$ in. $\times 2.54$ cm/in. ≈ 282 cm. Similarly, the upper estimate of 22 in. for a cubit leads to a height of 11 ft 9 in ≈ 358 cm. In modern times, there have been men over 9 ft tall, but not over 11 ft tall.

43. The weight W must satisfy $\text{BMI} = \dfrac{W}{h^2} = \dfrac{W}{1.90^2} < 25$, so $W < (1.90)^2 \times 25 = 90.25$ kg.

45. Answers will vary.

47. (a) If the species grew geometrically: Weight would scale as the cube of wingspan, so an individual with half the wingspan of an adult would have one-eighth the weight. However, the dimensions of this dinosaur probably did *not* grow geometrically; the wingspan probably grew not in proportion to the length of the dinosaur but more rapidly, so as to support the weight. Then an individual with half the wingspan would have one-fourth the wing area, so could support in flight only one-fourth the weight, or 25 lb.

(b) If the species grew geometrically: Weight would scale as the cube of wingspan, so an individual weighing half as much as an adult would have a wingspan $\sqrt[3]{\dfrac{1}{2}} \approx 0.79$ times as great. If instead the wingspan grew to support the weight: An individual weighing half as much as an adult would need half the wing area; with both length and width of the wing growing in the same proportion, wing area (and hence weight) would scale as the square of wingspan. Hence wingspan would scale as the square root of weight. The half-weight dinosaur would need a wingspan $\sqrt{\dfrac{1}{2}}$ times as large, or $50 \text{ ft} \times \sqrt{0.5} \approx 35$ ft.

49. Answers will also vary with assumptions about the height of Icarus. A 5-ft tall Icarus would have been about 15 times as long as a sparrow and hence had to fly $\sqrt{15} \times 20$ mph ≈ 77 mph. We assume that Icarus was a scaled-up sparrow, so that his wing loading was proportional to his length; with disproportionately large wings, the wing loading – and hence the minimum speed – would have been lower.

51. The square of wingspan is proportional to wing area, so the wing loading is proportional to weight divided by square of wingspan. For the 200-pounder, that ratio is $\frac{200}{50^2} = 0.080$, while for the 100-pounder it is $\frac{100}{36^2} \approx 0.077$. Close enough.

53. (a) (i) The giant ants are 8 m = 800 cm long, compared to the 1-cm length of a common ant. So the linear scaling factor is 800.

(ii) Since area scales as the square of the linear scaling factor, the surface area of the giant ant is $800^2 = 640{,}000$ times as large.

(iii) Since volume scales as the cube of the linear scaling factor, the volume of the giant ant is $800^3 = 512{,}000{,}000$ times as great.

(b) The giant ant has 800 times as much volume per unit of surface area, so its skin could supply one eight-hundredth of what it would need.

(c) There couldn't be any such giant ants.

55. Because $h \propto t^{1/4}$, we have $t \propto h^4$. For a tree to grow to 40 m tall, compared to a tree growing to 20 m tall, it will take $\left(\dfrac{40}{20}\right)^2 = 2^4 = 16$ times as long. So if it takes 30 years to grow to 20 m, it would take $16 \times 30 = 480$ years to grow to 40 m.

57. $A \propto d^2$ and $A \propto M^{3/4} \propto \left(d^2 h\right)^{3/4} = d^{3/2} h^{3/4}$, so $d^2 \propto d^{3/2} h^{3/4}$, hence $d^{1/2} \propto h^{3/4}$, and $d \propto h^{3/2}$.

59. A small warm-blooded animal has a large surface-area-to-volume ratio. Pound for pound, it loses heat more rapidly than a larger animal, hence must produce more heat per pound, resulting in a higher body temperature.

61. (a) On log-log paper:

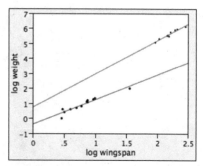

(b) Both relationships are allometric, since the results are good fits to straight lines whose slopes are not 1.

(c) The slope for birds is less steep than for planes.

Chapter 19
Symmetry and Patterns

Exercise Solutions

1. 5, 8 and 13

3. Answers will vary but will be Fibonacci numbers.

5. $1, 2, 3, 5, 8, \ldots, F_{n+1}$.

7. $m+n+\left(m+n\right)+\left(m+2n\right)+\left(2m+3n\right)+\left(3m+5n\right)+\left(5m+8n\right)$
    $+\left(8m+13n\right)+\left(13m+21n\right)+\left(21m+34n\right)=55m+88n,$

    eleven times the seventh number $5m+8n$.

9. (a) The digits after the decimal point do not change.
    (b) The digits after the decimal point do not change.
    (c) $\phi^2 = \phi + 1$
    (d) $\dfrac{1}{\phi} = \phi - 1$

11. (a) $\sqrt{3 \times 27} = \sqrt{81} = 9$
    (b) The area of the rectangle is $4 \times 64 = 256$, so the square must have side $\sqrt{256} = 16$.

13. $\sqrt[3]{3 \times 6 \times 12} = \sqrt[3]{216} = 6$

15. (a) 4, 7, 11, 18, 29, 47, 76, 123.
    (b) 3, 1.333, 1.75, 1.571, 1.636, 1.611, 1.621, 1.617, 1.618; The ratios approach ϕ.

17. Answers will vary. Equality holds exactly when $x = y$.

19. (a) 1, 1, 3, 5, 11, 21, 43, 85, 171, 341, 683, 1,365
    (b) $B_n = B_{n-1} + 2B_{n-2}$
    (c) 1, 3, 1.667, 2.2, 1.909, 2.048, 1.977, 2.012, 1.994, 2.003, 1.999
    (d) $x = 2, -1$; we discard the -1 root.; $B_n = \dfrac{2^n - \left(-1\right)^n}{3}$

21. Silver mean: $1 \pm \sqrt{2} \approx 2.414$; bronze mean: $\frac{1}{2}\left(3 \pm \sqrt{13}\right) \approx 3.303$;

    copper mean: $2 \pm \sqrt{5} \approx 4.236$; nickel mean: $\frac{1}{2}\left(5 \pm \sqrt{29}\right) \approx 5.193$.

    General expression: $\frac{1}{2}\left(m + \sqrt{m^2 + 4}\right)$.

23. All are true.

25. **(a)** B, C, D, E, H, I, K, O, X

(b) A, H, I, M, O, T, U, V, W, X, Y

(c) H, I, N, O, S, X, Z

27. **(a)** MOM, WOW; MUd and bUM reflect into each other, as do MOM and WOW.

(b) pod rotates into itself; MOM and WOW rotate into each other.

(c) Here are some possibilities: NOW NO; SWIMS; ON MON; CHECK BOOK BOX; OX HIDE.

29. For all parts, translations.

(a) Reflection in the horizontal midline of the C's .

(b) None other than translations

(c) Reflection in the horizontal midline, reflections in vertical lines through the centers of the H's or between them, 180° rotation around the centers of the H's or the midpoints between them, glide reflections.

(d) Reflection in vertical lines through the centers of the M's or between them.

31. *d5*

33. **(a)** *c5*

(b) *c12*

(c) *c22*

35. **(a)** *c6*

(b) *d2* (CBS)

(c) *d1* (Dodge Ram)

37. **(a)** *c4*

(b) *d2*

39. **(a)** *p1a1*

(b) *p1m1*

(c) *p111*

(d) *p112*

(e) *pm11*

(f) *pma2*

(g) *pmm2*

41. **(a)** *pmm2*

(b) *p1a1*

(c) *pma2*

(d) *p112*

(e) *pmm2* (perhaps)

(f) *p1m1*

(g) *pma2*

(h) *p111*

43. **(a)** Half-turns are preferred on Mesa Verde pottery, while reflections predominate on Begho smoking pipes.

(b) Neither culture completely excludes any strip type. Begho designs are heavily concentrated (almost 90%) in *p1m1*, *p112*, or *pmm2*, while Mesa Verde designs are more evenly distributed over the seven patterns.

(c) (i) *pm11* or *pma2*: Mesa Verde. (ii) *p112*: Mesa Verde. (iii) *pmm2*: Begho. (iv) *pm11*: Begho. (v) *p1m1*: Difficult to say. (vi) *pmm2*: Begho. (vii) *pmm2*: Begho. (viii) *pma2*: Mesa Verde. (ix) *p1a1*: Mesa Verde.

45. **(c)** Smallest rotation is 90°, there are reflections, there are reflections in lines that intersect at 45°: *p4m*.

(d) Smallest rotation is 90°, there are no reflections: *p4*.

47. Certainly none of the patterns with rotations at 60° or 120° can arise from this method, which eliminates the five patterns in the two bottom branches of the identification flowchart. For the vertical motion, four orientations are possible for the second triangle in the first column, corresponding to rotations clockwise by 0, 1, 2, or 3 times a rotation of 90° of the triangle in the upper left square around the center of its square. Similarly, for the horizontal motion, there are four choices, which we number in the same fashion. Then there are 16 combinations, which we can denote (with vertical motion first) as 00, 01, 02, 03, 04, 10, . . . , 44. However, some horizontal motions are not compatible with some vertical motions, and some of the patterns are reflections of others along the diagonal from upper left to lower right. In fact, the patterns *p4g* and *p4* cannot be realized because the original triangle has symmetry within itself, patterns *cm* and *cmm* cannot be realized because every square contains a triangle. The remaining eight can all be formed by the technique.

49. There are no knights facing up or down (so no rotations of 90°), no knights who are upside down (so no rotations of 180°, and no knights at 60° or 120° angles. So the smallest rotation is 360°, there are no reflections, but there is a glide reflection that takes yellow knights into brown knights. So the pattern is *pg*, provided color is disregarded; if not, then *p1*.

51. The block of three pentagons shaded in gray is repeated at rotations of 60° around the center of the "snowflake" outlined in red. There are no reflections, so the pattern is *p6*.

53. The intersection of the dashed lines are 120° rotation centers, but there are no reflection lines, so the pattern is *p3*

55. Answers will vary.

57. There is no identity element.

59. **(a)** *d2*.

(b) Any two of: *R* (180° rotation around the center), *V* (reflection in vertical line through its center), *H* (reflection in horizontal line through its center).

(c) {*I, R, V, H*}

61. Let *R* denote rotation counterclockwise by 90°, *V* reflection in vertical line through its center and *H* reflection in horizontal line through its center. Then the group can be written as $\{I, R, R^2, R^3, H, V, RH = VR, RV = HR\}$, where the last two elements are reflections across the diagonals.

63. Answers will vary. Here is one: $0 = (3 - 2) - 1 \neq 3 - (2 - 1) = 2$.

65. As in Example 6, number fixed positions, label with letters copies of the pattern elements in the positions, and pick a fixed position about which to make a half-turn R.

   (a) $\left\langle\, T, R \mid R^2 = I, \ T \circ R = R \circ T^{-1} \right\rangle = \left\{ \ldots, \ T^{-1}, \ I, \ T^{1}, \ \ldots; \ \ldots, \ R \circ T^{-1}, \ R, \ R \circ T^{1}, \ \ldots \right\}.$

   (b) $\left\langle T, R, H \mid R^2 = H^2 = I, \ T \circ H = H \circ T, \ R \circ H = H \circ R, \ \left(R \circ T\right)^2 = I \right\rangle =$

   $\left\{ \ldots, \ T^{-1}, \ I, \ T, \ \ldots; \ \ldots, \ R \circ T^{-1}, \ R, \ R \circ T, \ \ldots; \ \ldots, \ H \circ T^{-1}, \ H, \ H \circ T, \right.$

   $\left. \ldots; \ \ldots, R \circ H \circ T^{-1}, \ R \circ H, \ R \circ H \circ T, \ \ldots \right\}$

67. $\left\langle R \mid R^8 = I \right\rangle = \left\{ I, \ R, \ R^2, \ R^3, \ R^4, \ R^5, \ R^6, \ R^7 \right\}$, where R is a rotation by $45°$.

69. There are four rotation symmetries (including the identity), three reflection symmetries, and an inversion through the center.

71. The carved head is reproduced in the same shape at different scales.

73. & 75. Answers will vary.

Chapter 20
Tilings

Exercise Solutions

1. Exterior: 45°. Interior: 135°.

3. $180° - \dfrac{360°}{n}$.

5. The usual notation for a vertex type is to denote a regular n-gon by n, separate the sizes of polygons by periods, and list the polygons in clockwise order starting from the smallest number of sides, so that, e.g., 3.3.3.3.3.3 denotes six equilateral triangles meeting at a vertex. The possible vertex figures are 3.3.3.3.3.3, 3.3.3.3.6, 3.3.3.4.4, 3.3.4.3.4, 3.3.4.12, 3.4.3.12, 3.3.6.6, 3.6.3.6, 3.4.4.6, 3.4.6.4, 3.12.12, 4.4.4.4, 4.6.12, 4.8.8, 5.5.10, and 6.6.6.

7. 3.7.42, 3.9.18, 3.8.24, 3.10.15, and 4.5.20.

9. At each of the vertices except the center one, six triangles meet, with angles (in clockwise order) of 75°, 75°, 30°, 30°, 75°, and 75°.

11. Yes, because the half pentagon is a quadrilateral, and any quadrilateral can tile the plane.

13. (a) No.

    (b) No.

    (c) No.

15. Answers will vary.

17. No.; No.

19. The only way to tile by translations is to fit the outer "elbow" of one tile into the inner "elbow" of another. Labeling the corners as follows works: the corners on the top A and B, those on the rightmost side C and D, the middle of the bottom E, and the middle of the leftmost side F.

21. Just label the four corners consecutively A, B, C, and D.

23. Place the skew-tetromino on a coordinate system with unit length for the side of a square and with the lower left corner at $(0,0)$. Then $A = (1,2)$, $B = (3,2)$, $C = (2,0)$, and $D = (0,0)$ works.

25. Place the skew-tetromino on a coordinate system with unit length for the side of a square and with the lower left corner at $(0,0)$. Then $A = (0,1)$, $B = (1,2)$, $C = (3,2)$, $D = (3,1)$, $E = (2,0)$, and $F = (0,0)$ works.

26. If A and B coincide, so must D and E, and vice versa. Other such pairs: $B - C$ and $E - F$, and $C - D$ and $F - A$.

27. (a) Yes.

    (b) No.

    (c) No.

29. See figure below.

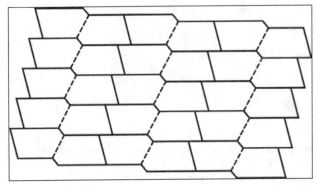

31. Answers will vary.

33. N, Z, W, P, y, I, L, V, X. See www.srcf.ucam.org/~jsm28/tiling/5-omino-trans.ps.gz.

35. Place the U on a coordinate system with unit length for the side of a square and with the lower left corner at $(0,0)$. Then $A = (2,2)$, $B = (3,2)$, $C = (3,0)$, $D = (1,0)$, $E = (0,0)$, and $F = (0,1)$ works. See www.srcf.ucam.org/~jsm28/tiling/5-omino-rot.ps.gz .

37. Answers will vary.

39. (a) Consecutive segments, all of the same length; each length gives a different tiling.

    (b) Consecutive repetition of any finite sequence of a's and b's.

41. (a) b; a; ab; aba; $abaab$; $abaababa$; $abaababaabaab$; $abaababaabaababaababa$

    (b) F_n segments at stage n

    (c) $2m + n$

43. *ABAABABA.*

45. The two leftmost A's would have had to come from two B's in a row in the preceding month.

47. Let S_n, A_n, and B_n be the total number of symbols, the number of A's, and the number of B's at the nth stage. We note that the only B's at the nth stage must have come from A's in the previous stage, so $B_n = A_{n-1}$. Similarly, the A's at the nth stage come from both A's and B's in the previous stage, so $A_n = A_{n-1} + B_{n-1}$. Using both of these facts together, we have $A_n = A_{n-1} + A_{n-2}$. We note that $A_1 = 0$, $A_2 = 1$, $A_3 = 1$, $A_4 = 2,\dots$. The A_n sequence obeys the same recurrence rule as the Fibonacci sequence and starts with the same values one step later; in fact, it is always just one step behind the Fibonacci sequence: $A_n = F_{n-1}$. Consequently, $B_n = A_{n-1} = F_{n-2}$, and $S_n = A_n + B_n = F_{n-1} + F_{n-2} = F_n$.

49. If a sequence ends in AA, its deflation ends in BB, which is impossible for a musical sequence. Similarly, if a sequence ends in $ABAB$, its deflation ends in AA, which we just showed to be impossible.

51. The first is, the second is not, part of a musical sequence: $ABAABABAAB \rightarrow ABAABA \rightarrow ABAB \rightarrow AA \rightarrow$ (2nd special rule) $BA \rightarrow$ (1st special rule) $ABA \rightarrow AB \rightarrow A$.

$ABAABABABA \rightarrow ABAAAB$, which has three A's.

53. If the sequence were periodic, the limiting ratio of A's to B's would be the same as the ratio in the repeating part, which would be a rational number, contrary to the result of Exercise 52.

Chapter 21
Savings Models

Exercise Solutions

1. The amount of interest per month you would pay Aunt Sally is $\dfrac{(\$30{,}000 \times 0.06)}{12} = \dfrac{\$1800}{12} = \$150.$
    For the nine months you would pay $9 \times \$150 = \$1350$ in (interest) payments.

3. **(a)** By the pattern shown, there is an increase by a factor if 2^n every $3n$ days. By doing some calculations, we can see that $2^{232} \approx 6.9 \times 10^{69}$ and $2^{233} \approx 1.4 \times 10^{70}$. So n is between 232 and 233. Thus, $3n$ is between 696 and 699. Calculating $2^{697/3}$ and $2^{698/3}$, we see that the better choice is 698 days.

    (b) We have $1000 = 10^3$. Since $\left(10^3\right)^{23} = 10^{69}$ and $\left(10^3\right)^{24} = 10^{72}$, an appropriate answer would be after 24 months.

5. **(a)** Since $A = P(1+rt)$, we have $A = \$1000(1+0.03 \times 1) = \$1000(1.03) = \$1030.00$. The annual yield is 3% since we have simple interest.

    (b) Since $A = P(1+i)^n$, $n = 1$, and $i = \frac{0.03}{1} = 0.03$, we have the following.

    $$A = \$1000(1+0.03) = \$1000(1.03) = \$1030.00$$

    The annual yield is 3% since we are dealing with simple interest because the money is compounded once during the period of one year.

    (c) Since $A = P(1+i)^n$, $n = 4$, and $i = \frac{0.03}{4}$, we have the following.

    $$\$1000 \times \left(1+\tfrac{0.03}{4}\right)^4 = \$1000(1.0075)^4 = \$1030.34$$

    Since $APY = \left(1+\dfrac{r}{m}\right)^m - 1$, we have $APY = \left(1+\dfrac{0.03}{4}\right)^4 - 1 \approx 3.034\%$.

    (d) $\$1000 \times \left(1+\frac{0.03}{365}\right)^{365} = \$1000(1.000082192)^{365} = \$1030.45$, with the same result for a 360-day or 366-day year.

    Since $APY = \left(1+\dfrac{r}{m}\right)^m - 1$, we have $APY = \left(1+\dfrac{0.03}{365}\right)^{365} - 1 \approx 3.045\%$, with the same result for a 360-day or 366-day year.

7. $APY = \left(1+\frac{r}{m}\right)^m - 1$, $m = 365$, and $\frac{r}{m} = \frac{0.0469}{365}$. We have $APY = \left(1+\frac{0.0469}{365}\right)^{365} - 1 \approx 4.801\%$.

9. $APY = \left(1+\frac{r}{m}\right)^m - 1$, $m = 2$, and $\frac{r}{m} = \frac{0.03}{2}$, we have $APY = \left(1+\frac{0.03}{2}\right)^2 - 1 \approx 3.023\%$. Since $(1.0323)^{20} \approx 1.89$, the bond will not double in value.

11. The interest is \$26.14 on a principal of \$7744.70, or $\frac{\$26.14}{\$7744.70} \times 100\% = 0.3375211435\%$ over 34 days. The daily interest rate is $\left(1.003375211435^{1/34} - 1\right) \times 100\% = 0.0099109\%$. The annual rate is then $\left(1.000099109^{365} - 1\right) \times 100\% = 3.684\%$.

13. **(a)** 3%: Predicted doubling time is $\dfrac{72}{100 \times 0.03} = \dfrac{72}{3} = 24$.

Since $A = P(1+i)^n$, $n = 24$, and $i = \frac{0.03}{1} = 0.03$, we have the following.
$$A = \$1000(1+0.03)^{24} = \$1000(1.03)^{24} = \$2032.79$$

4%: Predicted doubling time is $\dfrac{72}{100 \times 0.04} = \dfrac{72}{4} = 18$.

Since $A = P(1+i)^n$, $n = 18$, and $i = 0.04$, we have the following.
$$A = \$1000(1+0.04)^{18} = \$1000(1.04)^{18} = \$2025.82$$

6%: Predicted doubling time is $\dfrac{72}{100 \times 0.06} = \dfrac{72}{6} = 12$.

Since $A = P(1+i)^n$, $n = 12$, and $i = 0.06$, we have the following.
$$A = \$1000(1+0.06)^{12} = \$1000(1.06)^{12} = \$2012.20$$

(b) 8%: Predicted doubling time is $\dfrac{72}{100 \times 0.08} = \dfrac{72}{8} = 9$.

Since $A = P(1+i)^n$, $n = 9$, and $i = 0.08$, we have the following.
$$A = \$1000(1+0.08)^{9} = \$1000(1.08)^{9} = \$1999.00$$

9%: Predicted doubling time is $\dfrac{72}{100 \times 0.09} = \dfrac{72}{9} = 8$.

Since $A = P(1+i)^n$, $n = 8$, and $i = 0.09$, we have the following.
$$A = \$1000(1+0.09)^{8} = \$1000(1.09)^{8} = \$1992.56$$

(c) 12%: Predicted doubling time is $\dfrac{72}{100 \times 0.12} = \dfrac{72}{12} = 6$.

Since $A = P(1+i)^n$, $n = 6$, and $i = 0.12$, we have the following.
$$A = \$1000(1+0.12)^{6} = \$1000(1.12)^{6} = \$1973.82$$

24%: Predicted doubling time is $\dfrac{72}{100 \times 0.24} = \dfrac{72}{24} = 3$.

Since $A = P(1+i)^n$, $n = 3$, and $i = 0.24$, we have the following.
$$A = \$1000(1+0.24)^{3} = \$1000(1.24)^{3} = \$1906.62$$

36%: Predicted doubling time is $\dfrac{72}{100 \times 0.36} = \dfrac{72}{36} = 2$.

Since $A = P(1+i)^n$, $n = 2$, and $i = 0.36$, we have the following.
$$A = \$1000(1+0.36)^{2} = \$1000(1.36)^{2} = \$1849.60$$

(d) For small and intermediate interest rates, the rule of 72 gives good approximations to the doubling time.

15. **(a)** 2, 2.59, 2.705, 2.7169, 2.718280469

(b) 3, 6.19, 7.245, 7.3743, 7.389041321

(c) $e = 2.718281828\ldots$; $e^2 = 7.389056098\ldots$. Your calculator may give slightly different answers, because of its limited precision.

17. **(a)** Since $A = Pe^{rt}$ we have $A = \$1000e^{(0.03)(1)} = \$1000e^{0.03} = \$1030.45$. Thus, the interest is $\$1030.45 - \$1000.00 = \$30.45$.

(b) Since $A = P\left(1 + \frac{r}{365}\right)^{365}$ we have $A = \$1000\left(1 + \frac{0.03}{365}\right)^{365} = \$1030.45$. Thus, the interest is $\$1030.45 - \$1000.00 = \$30.45$.

In both cases, the interest is $\$30.45$, not taking into account any rounding to the nearest cent of the daily posted interest.

19. **(a)** We need to solve $(1+r)^2 = (1.00)(1.10)$. We have $(1+r)^2 = 1.10$, or $1.10^{1/2} = 1+r$, or $r = 1.10^{1/2} - 1 \approx 4.881\%$.

(b) We need to solve $e^{2r} = 1.10$. We have $2r = \ln 1.10$, or $r = \frac{1}{2}\ln 1.10 \approx 4.776\%$.

21. **(a)** $\left(e^{0.04} - 1\right) \times 100\% = 4.081\%$.

(b) The approximation for effective rate is $r + \frac{1}{2}r^2 = 0.05 + \frac{1}{2} \times (0.05)^2 = 0.05 + 0.00125 = 0.05125$ or 5.125%, very slightly less than the true effective rate.

23. Use the savings formula $A = d\left[\dfrac{(1+i)^n - 1}{i}\right]$, with $A = \$2000$, $i = \frac{0.05}{12}$, and $n = 24$.

$$\$2000 = d\left[\dfrac{\left(1 + \frac{0.05}{12}\right)^{24} - 1}{\frac{0.05}{12}}\right] \approx 25.18592053d$$

$$d = \dfrac{\$2000}{25.18592053} = \$79.41$$

25. Use the savings formula $A = d\left[\dfrac{(1+i)^n - 1}{i}\right]$, with $d = \$400$, $i = \frac{0.055}{12}$, and $n = 144$.

$$A = d\left[\dfrac{(1+i)^n - 1}{i}\right] = \$400\left[\dfrac{\left(1 + \frac{0.055}{12}\right)^{144} - 1}{\frac{0.055}{12}}\right] = \$81,327.45$$

27. Use the savings formula $A = d\left[\dfrac{(1+i)^n - 1}{i}\right]$, with $A = \$1,000,000$, $i = \frac{0.05}{12}$, and $n = 35 \times 12 = 420$.

$$\$1,000,000 = d\left[\dfrac{\left(1 + \frac{0.05}{12}\right)^{420} - 1}{\frac{0.05}{12}}\right] \approx 1136.092425d$$

$$d = \dfrac{\$1,000,000}{1136.092425} = \$880.21$$

29. Use the savings formula $A = d\left[\dfrac{(1+i)^n - 1}{i}\right]$, with $d = \$100$, $i = \frac{0.06}{12}$, and $n = 30 \times 12 = 360$.

$$A = d\left[\frac{(1+i)^n - 1}{i}\right] = \$100\left[\frac{\left(1 + \frac{0.06}{12}\right)^{360} - 1}{\frac{0.06}{12}}\right] = \$100{,}451.50$$

31. (a) $\dfrac{\$100}{1 - 0.32} = \$147.06$

(b) Use the savings formula $A = d\left[\dfrac{(1+i)^n - 1}{i}\right]$, with $d = \$147.06$, $i = \frac{0.075}{12}$, and

$n = 40 \times 12 = 480$ to calculate $A = \$147.06\left[\dfrac{\left(1 + \frac{0.075}{12}\right)^{480} - 1}{\frac{0.075}{12}}\right] = \$444{,}683.29$.

(c) $0.68 \times \$444{,}683.29 = \$302{,}384.64$

33. (a) $A = P(1.15)(1.07)(0.80) = 0.98440P$, so $r = (0.98440)^{1/3} - 1 = -0.00523 = -0.523\%$.

(b) It is the effective rate.

35. (a) exponential (decay)

(b) linear

(c) exponential (decay)

(d) linear

37. $\dfrac{\$10{,}000}{(1 + 0.05)^4} = \dfrac{\$10{,}000}{1.05^4} \approx \$8227.02$

39. $\dfrac{\$150{,}000}{(1 + 0.08)^{22}} = \dfrac{\$150{,}000}{1.08^{22}} \approx \$27{,}591.08$

41. (a) $\$(1.03)^3 = \$1.09$

(b) $\$1/1.09 = \$0.92$

43. $\$10{,}000 \times (1 - 0.12)^6 \times \left(\dfrac{1}{1 + 0.03}\right)^6 \approx \$3900$

45. Since $\dfrac{\text{cost in 2009}}{\$10.75} = \dfrac{\text{CPI for 2009}}{\text{CPI for 1962}} = \dfrac{227}{30.9} \approx 7.346278317$, we have the following.

$$\text{cost in 2009} = \$10.75(7.346278317) = \$78.97$$

Additional answers will vary.

47. Since $(1 + a)^{10} = 1 + \dfrac{207.2 - 160.5}{160.5}$, we have $(1 + a)^{10} = 1 + \dfrac{46.8}{160.5}$, or $(1 + a)^{10} = \dfrac{207.3}{160.5}$. Solving

for a, we have $1 + a = \sqrt[10]{\dfrac{207.3}{160.5}}$, or $a = \sqrt[10]{\dfrac{207.3}{160.5}} - 1 \approx 0.02592$ or 2.592%. Rounded to the nearest

whole percent, 3% is reasonable.

49. Let the purchasing power of the original salary be P. Then the purchasing power of the new

salary is $P \times 1.10 \times \dfrac{1}{1 + 0.20} \approx 0.91667P$, an 8.333% loss. Alternatively, we find

$\dfrac{0.1 - 0.2}{1 + 0.2} = \dfrac{-0.1}{1.2} \approx -0.08333$ or -8.333%.

51. Nowhere close

(a) As she ends her 35th year of service, her salary will be \$178,792.17, which we multiply by $\frac{1}{1.03^{35}}$ to get the equivalent in today's dollars: \$63,541.55. (We do not take into account that annual salaries are normally rounded to the nearest dollar or hundred dollars.) The result is easily obtained by use of a spreadsheet, proceeding through her salary year by year and then adjusting at the end for inflation.

$45,000	end of year
46800.00	2
48672.00	3
50618.88	4
52643.64	5
54749.38	6
58439.36	7
60776.93	8
63208.01	9
65736.33	10
68365.78	11
71100.41	12
75444.43	13
78462.21	14
81600.69	15
84864.72	16
88259.31	17
91789.68	18
95461.27	19

	end of year
99279.72	20
103250.91	21
107380.95	22
111676.18	23
116143.23	24
120788.96	25
125620.52	26
130645.34	27
135871.15	28
141306.00	29
146958.24	30
152836.57	31
158950.03	32
165308.03	33
171920.35	34
178797.17	35
	63541.55

(b) \$61,524.30

$61,524.30	end of year
47700.00	2
50562.00	3
53595.72	4
56811.46	5
60220.15	6
65333.36	7
69253.36	8
73408.56	9
77813.08	10
82481.86	11
87430.77	12
94176.62	13
99827.22	14
105816.85	15
112165.86	16
118895.81	17
126029.56	18
133591.34	19

	end of year
141606.82	20
150103.22	21
159109.42	22
168655.98	23
178775.34	24
189501.86	25
200871.97	26
212924.29	27
225699.75	28
239241.74	29
253596.24	30
268812.01	31
284940.73	32
302037.18	33
320159.41	34
339368.97	35
	61524.3

53. $\$2,000,000 \times \left(1 + x + \cdots + x^{19}\right)$, with $x = \frac{1}{1+a} = \frac{1}{1.03}$, giving $\$29.8$ million. If you can expect to earn interest rate r on funds once you receive them, through the last payment, then the present value of your stream of income of annual lottery payments P plus interest (with inflation rate a) is as follows.

$$P\left[\left(\frac{1+r}{1+a}\right)^{19} \cdot 1 + \left(\frac{1+r}{1+a}\right)^{18}\left(\frac{1}{1+a}\right)^{1} + \left(\frac{1+r}{1+a}\right)^{17}\left(\frac{1}{1+a}\right)^{2} + \cdots\right.$$

$$\left. \cdots + \left(\frac{1+r}{1+a}\right)^{1}\left(\frac{1}{1+a}\right)^{18} + 1 \cdot \left(\frac{1}{1+a}\right)^{19}\right] = P \frac{1}{(1+a)^{19}}\left[\frac{(1+r)^{20} - 1}{r}\right]$$

For $P = \$2$ million, $r = 4\%$, and $a = 3\%$, we get $\$33.4$ million. If you can earn 4% forever but inflation stays at 3%, the present value is infinite!

55. **(a)** Nestegg would be $\dfrac{\$100,000}{0.072} \approx \$1,388,888.89$. Monthly deposit would be as follows.

$$\$1,388,888.89\left[\frac{\frac{0.072}{12}}{\left(1 + \frac{0.072}{12}\right)^{420} - 1}\right] \approx \$735.16$$

(b) $\dfrac{\$100,000}{(1.03)^{35}} = \$35,538.34$

(c) $\dfrac{\$100,000}{(1.03)^{63}} = \$15,532.98$

57. Program the savings formula into the spreadsheet and vary the value of i until you find $A \geq \$5000$, using the Solver command in Excel, or otherwise: $i = 1.60\%$ per month, or an annual rate of $12 \times 1.60\% = 19.2\%$.

59. 4.97%. It is the effective rate.

Chapter 22
Borrowing Models

Exercise Solutions

1. The interest would be $\$5500 \times \dfrac{0.068}{12} \times 15 = \$467.50$. So you owe $\$5500 + \$467.50 = \$5967.50$

3. The first loan accumulates interest of $\$3500 \times \dfrac{0.068}{12} \times 51 = \$1011.50$. The second loan accumulates interest of $\$4500 \times \dfrac{0.068}{12} \times 39 = \$994.50$. The total amount of debt will be $\$3500 + \$1011.50 + \$4500 + \$994.50 = \$10,006.00$.

5. Since $A = P(1+i)^n$, $n = 5$, and $i = 0.06$, we have the following.
$$A = \$20,000(1+0.06)^5 = \$20,000(1.06)^5 \approx \$26,764.51$$

7. Since $A = P(1+i)^n$, $n = 30$, and $i = 0.06$, we have the following.
$$A = \$200,000(1+0.06)^{30} = \$200,000(1.06)^{30} \approx \$1,148,698.24$$

9. **(a)** The APR would be $365 \times 0.0005819 = 0.2123935$, or about 21.24%.

   (b) The effective annual rate would be $(1.0005819)^{365} - 1 \approx 0.2366$, or about 23.66%.

11. All but a tiny amount after 91 months

13. After 203 months (more than 16 years!), the balance is $500.16.

15. Assuming each month has 30 days, it will take 193 months (more than 16 years). You would pay more than $1800 interest.

17. Use the amortization payment formula $d = P\left[\dfrac{i}{1-(1+i)^{-n}}\right]$, with $P = \$14,462$, $i = \frac{0.029}{12}$, and $n = 36$.

$$d = P\left[\dfrac{i}{1-(1+i)^{-n}}\right] = \$14,462\left[\dfrac{\frac{0.029}{12}}{1-\left(1+\frac{0.029}{12}\right)^{-36}}\right] \approx \$419.94$$

19. Using the amortization payment formula $d = P\left[\dfrac{i}{1-\left(1+i\right)^{-n}}\right]$, with $P = \$14{,}462$, $i = \frac{0.049}{12}$, and $n = 60$ we have the following payment.

$$d = P\left[\frac{i}{1-\left(1+i\right)^{-n}}\right] = \$14{,}462\left[\frac{\frac{0.049}{12}}{1-\left(1+\frac{0.049}{12}\right)^{-60}}\right] \approx \$272.25$$

21. (a) Since $d = P\left[\dfrac{i}{1-\left(1+i\right)^{-n}}\right]$, we have $P = d\left[\dfrac{1-\left(1+i\right)^{-n}}{i}\right]$. with $d = \$170.75$, $i = \frac{0.073}{12}$, and $n = 84$ we have the following.

$$P = \$170.75\left[\frac{1-\left(1+\frac{0.073}{12}\right)^{-84}}{\frac{0.073}{12}}\right] \approx \$11{,}204.25$$

(b) You are making a total of $84 \times \$170.75 = \$14{,}343$ in payments. Since the cost of the car was $\$11{,}204.25$, $\$14{,}343 - \$11{,}204.25 = \$3138.75$ is paid in interest.

23. (a) Use the amortization payment formula $d = P\left[\dfrac{i}{1-\left(1+i\right)^{-n}}\right]$, with $P = \$20{,}000$, $i = \frac{0.068}{12}$, and $n = 12 \times 10 = 120$.

$$d = P\left[\frac{i}{1-\left(1+i\right)^{-n}}\right] = \$20{,}000\left[\frac{\frac{0.068}{12}}{1-\left(1+\frac{0.068}{12}\right)^{-120}}\right] \approx \$230.16$$

(b) You are making a total of $120 \times \$230.16 = \$27{,}619.20$ in payments. Thus, $\$27{,}619.20 - \$20{,}000.00 = \$7{,}619.20$ is paid in interest.

25. Use the amortization payment formula $d = P\left[\dfrac{i}{1-\left(1+i\right)^{-n}}\right]$, with $P = \$10{,}000$, $i = \frac{0.085}{12}$, and $n = 12 \times 10 = 120$.

$$d = P\left[\frac{i}{1-\left(1+i\right)^{-n}}\right] = \$10{,}000\left[\frac{\frac{0.085}{12}}{1-\left(1+\frac{0.085}{12}\right)^{-120}}\right] \approx \$123.99$$

27. (a) With the grace period, we accumulate interest for $45+6 = 51$ months. This is $\frac{51}{12} = 4.25$ years. Thus, giving us 17 quarters. Since $A = P\left(1+i\right)^{n}$, $P = \$10{,}471.20$, $n = 17$, and $i = \frac{0.0821}{4}$, we have to repay the following.

$$A = \$10{,}471.20\left(1+\tfrac{0.0821}{4}\right)^{17} \approx \$14{,}791.03$$

(b) Use the amortization payment formula $d = P\left[\dfrac{i}{1-\left(1+i\right)^{-n}}\right]$, with $P = \$14{,}791.03$, $i = \frac{0.0821}{12}$, and $n = 12 \times 20 = 240$.

$$d = P\left[\frac{i}{1-\left(1+i\right)^{-n}}\right] = \$14{,}791.03\left[\frac{\frac{0.0821}{12}}{1-\left(1+\frac{0.0821}{12}\right)^{-240}}\right] \approx \$125.66$$

(c) With 240 payments of $\$125.66$, you make $\$30{,}158.40$ in payments. The interest paid is $\$30{,}158.40 - \$10{,}471.20 = \$19{,}687.20$.

29. Use the amortization payment formula $d = P\left[\dfrac{i}{1-\left(1+i\right)^{-n}}\right]$, with $P = \$100{,}000$, $i = \dfrac{0.065}{12}$, and $n = 12 \times 30 = 360$.

$$d = P\left[\dfrac{i}{1-\left(1+i\right)^{-n}}\right] = \$100{,}000\left[\dfrac{\frac{0.065}{12}}{1-\left(1+\frac{0.065}{12}\right)^{-360}}\right] = \$632.07$$

31. Use the amortization payment formula $d = P\left[\dfrac{i}{1-\left(1+i\right)^{-n}}\right]$, with $P = \$100{,}000$, $i = \dfrac{0.06125}{12}$, and $n = 12 \times 15 = 180$.

$$d = P\left[\dfrac{i}{1-\left(1+i\right)^{-n}}\right] = \$100{,}000\left[\dfrac{\frac{0.06125}{12}}{1-\left(1+\frac{0.06125}{12}\right)^{-180}}\right] = \$850.62$$

33. Use the amortization formula $P = d\left[\dfrac{1-\left(1+i\right)^{-n}}{i}\right]$, with $d = \$632.07$, $i = \dfrac{0.065}{12}$, and $n = 360 - 12 \times 5 = 360 - 60 = 300$.

$$P = d\left[\dfrac{1-\left(1+i\right)^{-n}}{i}\right] = \$632.07\left[\dfrac{1-\left(1+\frac{0.065}{12}\right)^{-300}}{\frac{0.065}{12}}\right] = \$93{,}611.27$$

Thus, the amount of equity is $\$100{,}000 - \$93{,}611.27 = \$6388.73$.

35. Use the amortization formula $P = d\left[\dfrac{1-\left(1+i\right)^{-n}}{i}\right]$, with $d = \$850.62$, $i = \dfrac{0.06125}{12}$, and $n = 180 - 12 \times 5 = 180 - 60 = 120$.

$$A = d\left[\dfrac{1-\left(1+i\right)^{-n}}{i}\right] = \$850.62\left[\dfrac{1-\left(1+\frac{0.06125}{12}\right)^{-120}}{\frac{0.06125}{12}}\right] = \$76{,}186.80$$

Thus, the amount of equity is $\$100{,}000 - \$76{,}186.80 = \$23{,}813.20$.

37. Using Solver in Excel or otherwise, we get an annual rate of 3.68%.

39. If we consider that the interest accrues after every two weeks, then we solve the $\$15 = \$100 \times r \times \dfrac{2}{52}$ to find $r = 3.9 = 390\%$. If we consider that the interest accrues daily for 14 days, then we solve the $\$15 = \$100 \times r \times \dfrac{14}{365}$ to obtain $r \approx 3.91 = 391\%$.

41. The principal is $\$1500 - \$88 = \$1412$ and the (simple) interest over the week is \$88, so the interest rate is $100\% \times \frac{\$88}{\$1412} = 6.23\%$ for 1 week, for an annual rate of $52 \times 6.23\% = 324\%$. For a 365-day year, we get a daily rate of $\frac{6.23\%}{7} = 0.89\%$ per day and an annual percentage rate of $365 \times 0.89\% = 325\%$.

43. (a) Use the amortization payment formula $d = P\left[\dfrac{i}{1-(1+i)^{-n}}\right]$, with $P = \$100,000$, $i = \frac{0.08375}{12}$,

and $n = 12 \times 30 = 360$.

$$d = P\left[\frac{i}{1-(1+i)^{-n}}\right] = \$100,000\left[\frac{\frac{0.08375}{12}}{1-\left(1+\frac{0.08375}{12}\right)^{-360}}\right] = \$760.07$$

(b) Use the amortization payment formula $P = d\left[\dfrac{1-(1+i)^{-n}}{i}\right]$, with $d = \$760.07$, $i = \frac{0.08375}{12}$,

and $n = 360 - 12 \times 5 = 360 - 60 = 300$.

$$P = d\left[\frac{1-(1+i)^{-n}}{i}\right] = \$760.07\left[\frac{1-\left(1+\frac{0.08375}{12}\right)^{-300}}{\frac{0.08375}{12}}\right] = \$95,387.80$$

Thus, the amount of equity is $\$100,000 - \$95,387.80 = \$4612.20$.

(c) Use the amortization payment formula $d = P\left[\dfrac{i}{1-(1+i)^{-n}}\right]$, with $P = \$95,387.80$, $i = \frac{0.07}{12}$,

and $n = 12 \times 30 = 360$.

$$d = P\left[\frac{i}{1-(1+i)^{-n}}\right] = \$95,387.80\left[\frac{\frac{0.07}{12}}{1-\left(1+\frac{0.07}{12}\right)^{-360}}\right] = \$634.62$$

(d) Since the difference between the two payments is $\$760.07 - \$634.62 = \$125.45$, it would

take $\dfrac{\$2000}{\$125.45} \approx 15.9426$ or 16 months.

45. Given that he was to receive \$314.9 million in 30 payments, the annual payment would have

been $\dfrac{\$314.9}{30} \approx \$10.5$ million per year. Since one payment was to be received immediately, that

first payment accrues no interest. We have $\$314.9 - \$\left(\dfrac{314.9}{30}\right) = \$\left(\dfrac{9132.1}{30}\right) \approx \$304.4$ million to

be paid in 29 installments. $\$170 - \$\left(\dfrac{314.9}{30}\right) = \$\left(\dfrac{4785.1}{30}\right) \approx \$159.5$ million would be the

present value of the stream of payments. Using the amortization payment formula

$P = d\left[\dfrac{1-(1+i)^{-n}}{i}\right]$, with $d = \dfrac{\$314.9}{30}$, $P = \dfrac{\$4785.1}{30}$, and $n = 29$ we solve

$\dfrac{4785.1}{30} = \dfrac{314.9}{30}\left[\dfrac{1-(1+i)^{-29}}{i}\right]$. Notice we have $4785.1 = 314.9\left[\dfrac{1-(1+i)^{-29}}{i}\right]$ when we

multiply both sides of the equation by 30. We obtain $i \approx 4.97\%$. Answers can vary due to
rounding.

47. We assume payments at the end of the month.

For men: The Excel command = PV(0.04/12,199,2000,0,0) yields \$290,580.05.

Chapter 23
The Economics of Resources

Exercise Solutions

1. End of 2022 or beginning of 2023

middle of	China's population in millions	India's population in millions
2007	1318.00	1132.00
2008	1325.91	1150.11
2009	1333.86	1168.51
2010	1341.87	1187.21
2011	1349.92	1206.21
2012	1358.02	1225.50
2013	1366.17	1245.11
2014	1374.36	1265.03
2015	1382.61	1285.28
2016	1390.90	1305.84
2017	1399.25	1326.73
2018	1407.65	1347.96
2019	1416.09	1369.53
2020	1424.59	1391.44
2021	1433.14	1413.70
2022	1441.73	**1436.32**
2023	**1450.38**	**1459.30**

3. The population in mid-2025 is as follows.

    $$\left(\text{population in mid-2007}\right)\times\left(1+\text{growth rate}\right)^{18} = 4.086\left(1+0.017\right)^{18} \text{ billion} \approx 5.534 \text{ billion}$$

5. The population of Africa would be $925\times\left(1.024\right)^{18} = 1418$ million, almost 100% greater than, or twice as large, as Europe's population.

7. (a) $\dfrac{70}{0.5} = 140$ years

    (b) $\dfrac{70}{1.2} \approx 58$ years

9. (a) The population in mid-2025 is as follows.

    $$\left(\text{population in mid-2007}\right)\times\left(1+\text{growth rate}\right)^{18} = 6.625\left(1+0.012\right)^{18} \text{ billion} \approx 8.211 \text{ billion}$$

    The population in mid-2050 is as follows.

    $$\left(\text{population in mid-2007}\right)\times\left(1+\text{growth rate}\right)^{43} = 6.625\left(1+0.012\right)^{43} \text{ billion} \approx 11.065 \text{ billion}$$

    (b) No change in growth rate, no change in death rates, no global catastrophes, etc.

11. **(a)** The static reserve will be $\dfrac{2900}{84.7} \approx 34$ years.

    (b) The exponential reserve will be $\dfrac{\ln\left[1+\left(\frac{2900}{84.7}\right)(0.019)\right]}{\ln[1+0.019]} \approx 27$ years.

    (c) Answers will vary.

13. **(a)** The static reserve will be 100 years. We are seeking the exponential reserve. This will be

    $\dfrac{\ln\left[1+100(0.025)\right]}{\ln[1+0.025]} \approx 51$ years.

    (b) $\dfrac{\ln\left[1+1000(0.025)\right]}{\ln[1+0.025]} \approx 132$ years

    (c) $\dfrac{\ln\left[1+10{,}000(0.025)\right]}{\ln[1+0.025]} \approx 224$ years

15. **(a)** $\dfrac{\ln\left[1-100(0.005)\right]}{\ln[1-0.005]} \approx 138$ years

    (b) $\dfrac{\ln\left[1-100(0.01)\right]}{\ln[1-0.01]} = \dfrac{\ln(1-1)}{\ln(0.99)} = \dfrac{\ln 0}{\ln[0.99]}$

    This theoretically would imply forever!

17. $\dfrac{437 \times 10^{9} \text{ tons}}{100 \times 10^{6} \text{ plants} \times 800 \text{ years}} \approx 5.5$ tons/plant/year ≈ 30 lb/plant/day, which is unreasonable.

19. $\dfrac{1}{100}\ln\left(\dfrac{62.95}{3.93}\right) \approx 2.77\%$

21. After the first year, the population stays at 15.

23. 7, 18.2, 6.6, 17.6, 8.4, 19.5, 2.0, 7.3, 18.6, 5.3

25. We must have $f(x_n)=x_n$, or $4x_n(1-0.05x_n)=x_n$.   The only solutions are $x_n=0$ and $4(1-0.05x_n)=1$, or $x_n=15$.

27. Using $x_{n+1}=f(x_n)=3x_n(1-0.05x_n)$ with $x_1=10$ we have the following (rounded).

    10, 15.0, 11.3, 14.8, 11.6, 14.6, 11.8, 14.5, 11.9, 14.4

    The population is oscillating but slowly converging to $\frac{40}{3} \approx 13.3$.

29. We must have $f(x_n)=x_n$, or $3x_n(1-0.05x_n)=x_n$.   The only solutions are $x_n=0$ and $3(1-0.05x_n)=1$, or $x_n=\frac{40}{3}\approx 13.3$.

31. The red dashed line indicates the same size population next year as this year; where it intersects the blue curve is the equilibrium population size.

33. Using $x_{n+1} = f(x_n) = 1.5x_n(1 - 0.025x_n)$ with $x_1 = 11$ we have the following (rounded).

$$11, 12.0, 12.6, 12.9, 13.1, 13.2, 13.3, 13.3, 13.3, 13.3$$

35. The population sizes are 11, 15.0, 13.7, 14.9, 14.9, 15.0, 14.8, 14.3, 13.0, 9.5 – and the following year the population is wiped out.

37. About 15 million pounds. Maximum sustainable yield is about 35 million pounds for an initial population of 25 million pounds.

39. **(a)** The last entry shown for the first sequence is the fourth entry of the second sequence, so the first "joins" the second and they then both end up going through the same cycle (loop) of numbers over and over.

  (b) 39, 78, 56, and we have "joined" the second sequence. However, an initial 00 stays 00 forever; and any other initial number ending in 0 "joins" the loop sequence 20, 40, 80, 60, 20, ...

  (c) Regardless of the original number, after the second push of the key we have a number divisible by 4, and all subsequent numbers are divisible by 4. There are 25 such numbers between 00 and 99. You can verify that an initial number either joins the self-loop 00 (the only such numbers are 00, 50, and 25); joins the loop 20, 40, 80, 60, 20, . . . (the only such are the multiples of 5 other than 00, 50 and 25); or joins the big loop of the other 20 multiples of 4.

41. **(a)** 133, 19, 82, 68, 100, 1, 1, ... The sequence stabilizes at 1.

  (b) Answers will vary.

  (c) That would trivialize the exercise!

  (d) For simplicity, limit consideration to 3-digit numbers. Then the largest value of f for any 3-digit number is $9^2 + 9^2 + 9^2 = 243$. For numbers between 1 and 243, the largest value of f is $1^2 + 9^2 + 9^2 = 163$. Thus, if we iterate f over and over – say 164 times – starting with any number between 1 and 163, we must eventually repeat a number, since there are only 163 potentially different results. And once a number repeats, we have a cycle. Thus, applying f to any 3-digit number eventually produces a cycle. How many different cycles are there? That we leave you to work out.

  Hints: 1) There aren't very many cycles.

  2) There is symmetry in the problem, in that some pairs of numbers give the same result; for example, $f(68) = f(86)$.

43. **(a)** 0.0397, 0.15407173, 0.545072626, 1.288978, 0.171519142, 0.59782012, 1.31911379, 0.0562715776, 0.215586839, **0.722914301**, 1.32384194, 0.0376952973, 0.146518383, 0.521670621, 1.27026177, 0.240352173, 0.78810119, 1.2890943, 0.171084847, **0.596529312**

  (b) **0.723**, 1.323813, 0.0378094231, 0.146949035, 0.523014083, 1.27142514, 0.236134903, 0.777260536, 1.29664032, 0.142732915, **0.509813606**

  (c) **0.722**, 1.324148, 0.0364882223, 0.141958718, 0.507378039, 1.25721473, 0.287092278, 0.901103183, 1.16845189, **0.577968093**

45. Period 2 begins at $\lambda = 3$, period 4 at $1 + \sqrt{6} \approx 3.449$, period 8 at 3.544, period 3 at $1 + 2\sqrt{2} \approx 3.828$, and chaotic behavior onsets at about 3.57.

See http://www.answers.com/topic/logistic-map .